CW00322804

Models of Opportunity

Entrepreneurship is changing. Technology and social networks create a smaller world, but widen the opportunity horizon. Today's entrepreneurs build organizations and create value in entirely new ways and with entirely new tools. Rather than just exploit new ideas, innovative entrepreneurs design organizations to make sense of unlikely opportunities. The time has come to overhaul what we know about entrepreneurship and business models. *Models of Opportunity* links scholarly research on business models and organizational design to the reality of building entrepreneurial firms. It provides actionable advice based on a deeper understanding of how business models function and change. The six insights that form the core of the book extend corporate strategy and entrepreneurship in a completely new direction. Case studies of innovative companies across industries demonstrate how visionary entrepreneurs achieve unexpected results. The insights, tools, and cases provide a fresh perspective on emerging trends in entrepreneurship, organizational change, and high-growth firms.

GERARD GEORGE is Professor and Deputy Head of Innovation and Entrepreneurship at Imperial College London, where he also serves as Director of the Rajiv Gandhi Centre facilitating its strategic commitments in India for joint research, technology commercialization, and educational programs. He is Professorial Fellow of the UK's Economic and Social Research Council and Associate Editor of the *Academy of Management Journal*. Previously, he was a tenured professor at London Business School and at the University of Wisconsin-Madison. He serves on the boards of several high-technology companies and is actively engaged in guiding numerous start-ups and large companies on innovation, venturing, and entrepreneurship.

ADAM J. BOCK is Lecturer in Entrepreneurship at the University of Edinburgh Business School. He has published research in *Entrepreneurship Theory and Practice* and *Journal of Management Studies*. He is also the co-author of *Entrepreneurship in the Research Context* (2009), a web-based e-learning course for nascent scientist entrepreneurs. Adam is an experienced entrepreneur and venture financier. He co-founded three medical device companies and managed three angel investor networks in the United States, facilitating more than $10 million of investment into start-up companies.

Models of Opportunity

How entrepreneurs design firms to achieve the unexpected

GERARD GEORGE

AND

ADAM J. BOCK

CAMBRIDGE UNIVERSITY PRESS
Cambridge, New York, Melbourne, Madrid, Cape Town,
Singapore, São Paulo, Delhi, Tokyo, Mexico City

Cambridge University Press
The Edinburgh Building, Cambridge CB2 8RU, UK

Published in the United States of America by
Cambridge University Press, New York

www.cambridge.org
Information on this title: www.cambridge.org/9780521170840

First published 2012

Printed in the United Kingdom at the University Press, Cambridge

A catalog record for this publication is available from the British Library

Library of Congress Cataloguing in Publication data
George, Gerard.
Models of opportunity : how entrepreneurs design firms to
achieve the unexpected / Gerard George, Adam J. Bock.
pages cm
Includes bibliographical references and index.
ISBN 978-0-521-76507-7
1. Entrepreneurship. 2. Organizational change. 3. Technological
innovations. I. Bock, Adam J. II. Title.
HB615.G465 2012
658.4'21–dc23
201104865

ISBN 978-0-521-76507-7 Hardback
ISBN 978-0-521-17084-0 Paperback

To Hema, Vivian, and Maegan, who inspire me to achieve the unexpected.

GG

To my own little hopeful monsters, Taran Lee and Kenna Rose.

AJB

Contents

Figures

Tables

Boxes

Acknowledgments

Few books write themselves. This book represented more work, but also more discovery and excitement than we ever anticipated. It also required participation and support from many dedicated, smart, and busy people.

To start, we thank Professor Massimo Warglien of the Università Ca'Foscari di Venezia (Venice). His thoughts and ideas helped us develop and elaborate the central themes of the book. He wrote the software code for the coherence simulation that anchors Chapter 3, and was inspirational in directing us to better understand the importance of organizational narratives.

We also thank Paula Parish and her team at Cambridge University Press, who patiently waited, and waited, and waited. Their assistance in coordinating, delivering, and marketing this book has proved invaluable.

The illustrations at each chapter header, which helped us visualize the book's core narrative, are the work of Anirudha Venkata Surabhi. We gratefully acknowledge and thank his creative inspiration.

The financial support of the UK's Economic and Social Research Council (ESRC) through Gerry's Professorial Fellowship (RES-051-27-0321) and the Advanced Institute of Management Innovation Fellowship (RES-331-27-0011) was instrumental in allowing him to dedicate time, effort and resources to this project.

We are indebted to the founders, executives, managers, and line staffers at all of the firms that participated in our research. Hundreds of people put aside the day-to-day needs of their organizations, despite incredible pressures and challenges, to share their time, knowledge, and wisdom. They did this in the full knowledge that their participation carried no direct personal benefit to them. They told us their stories, shared their tragedies and triumphs, and answered our odd questions. They were, to a person, interested, dedicated, and conscientious. Some of these individuals have waited more than four years to see what on earth we would write about them.

The vast majority of the people who helped us aren't mentioned in the book, but their contribution to our learning is not diminished by that. And a few companies that participated in our studies are not mentioned at all. This is an unfortunate reality of book-writing: the process of creating a coherent narrative benefits from simplicity and focus. Each company's story could have been a book on its own, but our purpose required the application of Occam's razor far more often than we would have liked.

Some of the data on business model innovation was graciously provided by IBM's Institute of Business Value. We thank Dr. Stephen Ballou and his team for their patience and generosity in providing access to this valuable data.

We also express our gratitude to Imperial College London and the University of Edinburgh for providing the institutional support to complete this research. Particularly, our most sincere thanks to Anushka Patel and Catherine Appleton who have helped us channel our joint efforts in this book.

Finally, but by no means the least in the order of importance, we once again thank our respective wives, and children, family, friends, and colleagues for their tolerance and encouragement, without which this book would not have reached fruition.

Six insights to achieve the unexpected

Entrepreneurship is changing. Much of what we know about opportunities and venture management is changing as well. Opportunities dismissed as implausible or even impossible have become reality; in some cases commonplace and indispensable. Thousands, perhaps millions of entrepreneurs strive every day to achieve the unexpected by transforming implausible ideas into viable companies. What distinguishes the phenomenal from the futile?

Consider the following implausible ideas:

- Serving 200 million wireless customers with no infrastructure
- testing drug safety on beating heart cells grown in a petri dish
- sorting the world's information using heuristics based on academic paper citation models
- educating low-income, urban American children at publicly funded boarding schools

- providing free videoconferencing worldwide
- stopping spam email by identifying reliable senders rather than fake content
- creating a $40 artificial knee
- auctioning anything at anytime for anyone.

Some of these opportunities, once unimaginable, have now become mundane: eBay, Facebook, Google, and Skype being the most obvious. The $40 'JaipurKnee' project originally developed by students at Stanford University is being fitted to patients in India, bringing hope to leg amputee patients worldwide. The others are active, well-known businesses, if perhaps described in unfamiliar ways. You might not have recognized Bharti Airtel (the Indian mobile services company), Return Path (the safe sender email firm), The SEED Foundation (creator of the only urban, public boarding schools in the United States), or Cellular Dynamics (the stem cell company). Most of us would have dismissed these ideas as strange or impossible – yet, some enterprising individuals see the world differently. Transforming implausible ideas into plausible opportunities presents a fundamentally new form, or perhaps even art, of entrepreneurship.

SHRINKING WORLD, EXPANDING OPPORTUNITIES

The global economy has become a cliché. Outsourcing, transnational distribution, and the interconnectivity of financial markets are familiar elements of the "smaller world" theme. High-profile entrepreneurship leads blog and newspaper headlines, especially in topical areas of renewable energy, internet infrastructure, and mobile communications sectors. But the vast majority of nascent entrepreneurship remains mostly ignored. Entrepreneurial activity, at the identification and enactment stage in local contexts, is where smaller world effects have the greatest impact. Every entrepreneurship story, great or small, starts somewhere. We need to update our understanding of where, how, and why entrepreneurial stories are written. But first, we have to appreciate the drivers that have made these stories possible.

The unprecedented, global, exponential, and accelerating increase in access to information is unmistakable, creating new managerial challenges.[1] More entrepreneurs have access to more opportunities; the true scale of global entrepreneurial activity is just beginning to emerge.[2] The Kauffmann Foundation, an American non-profit organization that champions the research and practice of entrepreneurship, estimates that roughly 4 percent of the US population participates in entrepreneurial activity each year. This is consistent with government estimates indicating 10–15 million active entrepreneurs in the United States. And while much has been made of the rise of *consumer* markets in emerging economies in Asia, it is the explosion in entrepreneurial activity that should really capture our attention.

If entrepreneurial activity is only half as prevalent in India and China, the base population of entrepreneurs easily exceeds 50 million individuals. Other estimates, however, suggest that the rapid development of low-income populations in these countries is leading to *higher* rates of entrepreneurial activity. The world's entrepreneurial community may soon exceed 100 million, or even 200 million individuals.[3] While these entrepreneurs lack some of the advantages enjoyed in relatively resource-rich economies like the United States, much less hotspots like Silicon Valley, the democratization of information resources and international networks strongly suggests that gaps in resource access will contract rather than expand. In other words, the coming decades will see a phenomenon never previously possible: a ten- or hundred-fold increase in the number of active, capable entrepreneurs in the global economy.[4]

While the scope and scale of potential opportunity has expanded, the geography of exploiting those opportunities has dramatically contracted. For the last few decades, Western businesses off-shored low-cost manufacturing to rapidly achieve scale economies and facilitate geographically dispersed product distribution. But the new world of opportunities includes instantaneous access to global markets: anyone anywhere can upload an application to the iTunes™ App Store to reach users worldwide. The "small world" effect goes far beyond logistics and distribution. It includes the potential for access to every type of human, physical, and intellectual property resource, often with nearly zero marginal cost for coordination.

Similarly, venturing mechanisms available to entrepreneurs have dramatically expanded. Sophisticated financing, legal, and operational structures have emerged to fund and grow ventures across industries and geographies. Consider the case of entrepreneur Cory Cullinan, who gave up a stable and rewarding music educator job to become Doctor Noize, the creator of an edutainment world inhabited by musical creatures. His award-winning album, *The Ballad of Phineas McBoof* led to a financing partnership with Hong Kong-based developer Outblaze and its multimedia production company Dream Cortex.[5] Cory provides the creative firepower while Dave Kim of Outblaze provides the operational oversight to coordinate designers, engineers, and developers in three countries. Doctor Noize has a SiriusKids™ #1 hit song ("Bananas"), also successfully released as an mobile app via the iTunes App Store. The second album will be released this fall, along with a second children's book. Doctor Noize has been commissioned to write a symphony with California's North State Symphony.[6] Geographical, organizational, and structural barriers to complex models of entrepreneurial action are rapidly disappearing.

Although much entrepreneurship is driven by economic need, entrepreneurs have more, and more unusual options for adopting or even creating new roles in opportunity development. Successful entrepreneurs employ a complex portfolio of skills and talents to identify and exploit opportunities. In the past, tacit knowledge specialization restricted activity to entrepreneurs' base of experience or education. In a world enabled by instantly accessible networks, entrepreneurs may leverage a variety of roles, identities, and organizational structuring options to create highly idiosyncratic personal journeys that facilitate, and are facilitated by, a commercial opportunity. The popular media prefers to focus on the solo entrepreneur struggling against faceless, monolithic corporate giants, but this is only one narrative entrepreneurs exploit to further creative endeavors. Return Path CEO Matt Blumberg is quick to point out that most great start-ups have more than one founder, and Return Path is no exception. As we'll discuss later, entrepreneurs achieve the unexpected by creating coherent narratives. But these narratives sometimes blur or obscure the complexity of entrepreneurial action. Return Path's narrative derives from the experience of multiple founders, including company president George Bilbrey and CFO Jack Sinclair, as well as corporate transactions such as acquisitions, divestitures, and financing events.

Some entrepreneurs reach beyond the private sector to coordinate resources and structures with non-commercial entities such as universities, governments, and charities. At the University of Wisconsin-Madison, Professor Miron Livny runs Project Condor, which provides a high-throughput computing platform to a variety of customers including universities, governments, and Fortune 500 companies. The figure shown below maps most of the pools of computers available on the Condor system.[7] The graphic provides an excellent example of how entrepreneurs have both enabled and benefited from a smaller, interconnected world. The reach of this small *organization*, which could easily be mistaken for a software company, has global impact, but operates entirely within the University as a funded research project. This model of opportunity didn't become physically operable until ten years ago and was highly implausible when it was started 20 years ago. Much before that, it wasn't even conceivable.

Many entrepreneurial ventures are now "born global."[8] Even when multinational distribution and sales costs are non-trivial, establishing global operations has become possible, expected, or necessary in some sectors. Rather than traditional expansion models in which

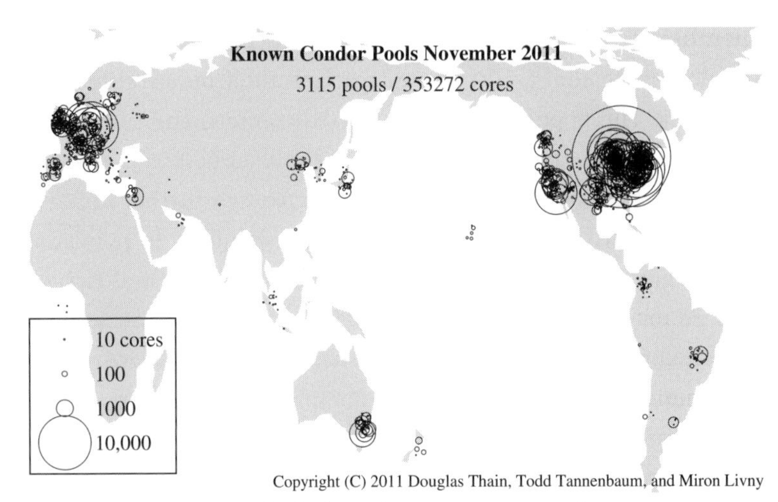

FIGURE Condor high-throughput computing pools[9]

firms grow locally, regionally, and nationally before extending into international markets, "born global" ventures develop international presence parallel to or even in advance of home country presence. We tend to think of "multinational firms" as large, familiar corporations. But it is entrepreneurial firms that lead the way in creating truly transnational organizations whose headquarters location is largely one of convenience, facilitated by inexpensive telecommunications, reliable information technology systems, and transparent regulations and legal structures.

Equally important, entrepreneurs can realize global impact, fueled by online social networks and global news sourcing and distribution, often in real time, across a variety of information platforms. In the past, entrepreneurs needed to leverage the legitimacy of established organizations and build on perfectly consistent internal systems to reshape industries.[10] But in the late 1990s and 2000s, extraordinary entrepreneurs and firms authored entirely new industries by expanding the boundaries of what seemed plausible. Within 36 months of its founding in 2003, Skype was responsible for more than 25 *billion* minutes of international voice communications, representing 4.4 percent of the global voice communications market.

36 months later, that share had tripled to 13 percent. In 2011 the company recorded an instance of 30 million *simultaneous* users, as if nearly the entire population of Canada were all using the same communication system at the same time.[11]

Just as the origins and journeys of entrepreneurs have changed, so too entrepreneurial exits have taken on entirely new meaning. We now distinguish between firm exits, personal exits, technology exits, and financial exits. In the IPO-rich decade of the 1990s, entrepreneurial exit was synonymous with shareholder exit, when funders monetized their investments, usually through acquisition or IPO.[12] But for most entrepreneurs, the monetization exit is not equivalent to personal exit. Many entrepreneurs experience multiple exits as specific roles conclude. Professor Franco Cerrina was one of the three scientific founders of Nimblegen, a life science company that developed novel DNA testing technology. He first phased out of active technology development at the company to return to a full-time university position. He later ended his executive advisory role, and finally cashed out his shareholder interest when the company was acquired by Roche in 2007.[13] Entrepreneurial journeys have become increasingly complex, fascinating, strange, and wonderful.[14]

REVISITING ENTREPRENEURSHIP

The authors of this book are both scholars and practitioners of entrepreneurship. We have realized that some entrepreneurial journeys simply don't fit within traditional management rules and theories. Our mental models about entrepreneurs need updating. While the mainstream media focuses on seemingly new incarnations of Andrew Carnegie, Steve Jobs, Ingvar Kamprad, or Jamsetji Tata, most dramas are unfolding on entirely different stages. The vast majority of novel commercial activity will arise in entirely new contexts, driven by a cohort of entrepreneurs who perceive the world very differently than twentieth-century industrialists. The commercial exploitation of exciting opportunities is evolving to new criteria, and changing those criteria in the process. The popular conceptualization of entrepreneurship as personal wealth-creation via the means of company-building doesn't explain a growing cadre of entrepreneurs tackling unusual, important, and implausible opportunities. Entrepreneurship is no longer just about making money; it is much more about making opportunities *real.*

Entrepreneurship textbooks and business school courses often focus on opportunity assessment, fit with environment, competitive advantage, financing, and growth.[15] These are, to be sure, important and relevant aspects of venturing and firm growth. But our experience, observations, and research suggest that this perspective has become narrow and outdated. Entrepreneurs may use these traditional managerial methods as tools, but they do so within a richer, kaleidoscopic confluence of possibilities. Endearing flaws, coherence, and narrative may seem like unlikely organizational qualities, but when firms enact implausible opportunities, the tools they use may seem implausible as well.

Now is the time to study the changing nature of opportunities and entrepreneurship itself. There will always be a place for investigating drivers of growth, profitability, and economic impact. But right now and for the next few decades perhaps, there are even more interesting questions to ask. The answers to these questions will affect hundreds of millions of people involved in this new world of entrepreneurial activity. How do entrepreneurs shape the very opportunities they exploit? How are their entrepreneurial journeys shaped by the novel challenges they face? How do entrepreneurs design organizations when the tenets of organizational design lag rapidly changing technological and social capabilities?

We present a new framework for thinking about entrepreneurs. Not as heroes, though they may be heroic. Not as managers, though they must apply managerial skills. Not as risk-takers, innovators, or explorers, though many fulfill these roles as well. *We see entrepreneurs as authors writing organizational stories as they happen.* Entrepreneurs design organizations to realize opportunities and build bridges to span opportunity gaps. But ultimately the most unexpected and fantastic results emerge when entrepreneurs narrate their own story even as they make it real.

This book does not attempt to answer who *is* or *becomes* an entrepreneur. Instead we observe and recount changes in how entrepreneurs address opportunities under significant uncertainty. This is, perhaps, an imperfect research process, in that we cannot build our story on firmly established first principles. An analogy may help explain our intent. We might like to know exactly what protons, neutrons, and quarks really are, but we can still investigate the broader properties of cabbages, kings, and start-ups while we wait. For simplicity, we therefore return to one of the earliest definitions of entrepreneurship as a basis for discussion. More than a century of research on management and entrepreneurship hasn't established a better foundation for thinking about entrepreneurship than the iconic work of Joseph Schumpeter, the Austrian economist and finance minister. One hundred years ago Schumpeter argued that entrepreneurs invest time and other resources to create "new combinations" in a process he coined "creative destruction."[16] This then, is the foundation of the entrepreneurial process, which we define broadly as *realizing new opportunities through the design of organization.*

This doesn't imply that solo business owners are not entrepreneurs – there is organization present when even one person trades. But this does exclude, carefully and directly, the *inventor* who, for whatever reason, is not involved in how that innovation is ever used. This is an important distinction between *opportunity* and *innovation*. For example, while Thomas Edison's experiments led to a viable light bulb, his efforts to link the innovation to centralized electrical energy production and infrastructure were equally important. The light bulb was the *innovation*; residential and commercial electrical lighting was the *opportunity.*

Our framework also excludes organizations, large and small, pursuing profits or other goals associated with exploiting *existing* opportunities.

Defining an "opportunity" presents similar challenges. Again, while we do not presume to resolve the scholarly debate, we want to clarify how we will discuss opportunities in this book. One perspective argues that opportunities are static mechanisms for generating profits. Within that paradigm, an effectively infinite set of opportunities exists *in potential.* Some have been exploited; some will be exploited; and some will remain forever unrealized.[17] When the right person at the right moment has the right insight, a specific opportunity becomes accessible.

An alternate framework for thinking about entrepreneurship argues that opportunities do not exist in a sort of ethereal waiting room. It is the exploitation activities of the entrepreneur that define and make the opportunity real.[18] This distinction may seem subtle, but it is important to the approach we take in this book. We believe that opportunities are not just identified by alert observation, or even primarily accessed via prior knowledge. *The active, dynamic, and iterative activity of the entrepreneur reveals and then shapes the opportunity, even as the entrepreneur is shaped by that very process.*

With these foundations in mind, our research has helped reveal three important characteristics of the process by which entrepreneurs enact and exploit opportunities:

First, realizing an opportunity is not a linear, step-wise, or optimizing process. Broad trends and key factors may be statistically or anecdotally linked to success or failure, but it is not possible to predict with certainty whether opportunities, especially *interesting* opportunities, can or will be successfully commercialized.

Second, opportunities are plastic, not rigid. The nature and realization of any opportunity is at least partly contingent on circumstances and initial conditions, and probably specific to the individual who identifies it, but opportunities change and evolve, sometimes in unpredictable ways.

Third, and perhaps most important, the entrepreneurial journey itself changes both the opportunity and the entrepreneur. This dynamic and inherently uncertain process emerges from the interaction between the entrepreneur, the opportunity, the exploitation path, and an often rapidly changing environment. As we have seen in our prior research, entrepreneurship is a transformative experience; this is especially true when the opportunity is improbable and the process fluid.

HOW WE GOT HERE

This book presents the first steps, hesitant though enthusiastic, towards a new understanding of entrepreneurial activity. As such, we rely on the results of prior research, our own direct observations and experience, and some speculative creativity. We have tried to balance the rigor of our academic training with the risk-taking and enthusiasm of our entrepreneurial experience.

The foundation of this story stems from research on academic scientists who commercialize their own inventions. Our previous book, *Inventing Entrepreneurs*, recounted the journeys of technology entrepreneurs and their firms. We documented the birth and growth of organizations and the identity transformation of the scientific founders. We began to perceive a vision of entrepreneurial activity at odds with established theory and popular beliefs. We heard entrepreneurs talking about "business models," but not as described in scholarly research. Business models had something to do with opportunities, and something to do with organizations, but consensus and clarity was nowhere to be found.

We know, now, that business models are at the heart of entrepreneurship, but at the time we found much disagreement, despite nearly ubiquitous reference, on how business models are used and even what they are. We undertook a comprehensive assessment of business models in both research and practice frameworks. This process included four separate but linked research studies. First, we conducted a pilot study to understand how managers working in entrepreneurial spaces understood business models. We then reviewed more than 500 published research papers to survey the field. We realized that researchers and entrepreneurs either used different words, or the same words in different ways, to explain and discuss business models. This, then, was the first step in our journey: to find a way to talk about business models that would reflect not only how scholars study them but also how entrepreneurs use them.

We initiated a detailed survey of more than 200 managers of small and medium-sized businesses in India and the UK. We learned that the language of business models links opportunities to organizational structures. This is important for three reasons. First, it integrated prior research on business models, even if that required reinterpreting some assumptions and findings. Second, it ensured that we could more effectively talk to entrepreneurs about opportunities and organizations and make sense of their experiences. Third, it suggested a bridge for how entrepreneurs develop and adapt business models in complex environments.

But entrepreneurship and opportunity exploitation is not limited to small and medium-sized firms. To better understand entrepreneurial activity on a global basis, we needed global data. We are grateful to IBM's Institute for Business Value for providing access to interviews with CEOs of more than 750 large organizations worldwide. Business model innovation is not limited to new ventures. The identification and exploitation of new opportunities appears to be more important to large firms than ever before, and may be shaping the competitive landscape in new and unfamiliar ways.

Finally, we conducted more than 200 hours of interviews and observations at 11 entrepreneurial firms in the US, UK, and India. We chose these firms not based on expectations of success or failure, but to present a spectrum of entrepreneurial opportunities across industries, countries, and development stages. Our one bias was to identify organizations developing *interesting* opportunities. To the extent that these therefore represent outliers, we can only point out that the organizations that most influence our lives, such as Wal-Mart, Tata, Toyota, the United Nations, Apple, Microsoft, ADM,

Cisco, Merck, and so on, were all once entrepreneurial ventures in one form or another that achieved tremendous success, and are therefore, *by definition*, outliers. We make no explicit predictions as to whether the companies we studied will be household names in the future, though the odds are against it. Rather, we wanted to show the uniqueness inherent to most organizations, and the strange, predominantly unstudied process of entrepreneurial activity that, in a few cases, changes the world. We reconstructed the telling of entrepreneurial narrative at the heart of these firms, witnessing the triumph and tragedy inherent to the pursuit of implausible opportunities.

In the end, we realized that clarifying the language of opportunities and business models was only half of the story: a symptom of the more fundamental disconnect. The lingua franca of organizations is dominated by the mindset of strategy and organizational behavior. As such, it treats opportunities as static and exogenous, focused on accumulating, adapting, and fitting resources to convert opportunity into profit. More recent works in strategy emphasize resilience[19] and temporary advantage over sustained competitive advantage.[20] These new research directions on organizational behavior and strategy have begun to incorporate aspects of entrepreneurial existence, but still must operate within a context driven by traditional views about industrial communities and competition. The emerging generation of entrepreneurs addresses the evolving global landscape with new beliefs and ideas. Traditional perspectives on how organizations *succeed* need to reflect how entrepreneurs make unlikely opportunities *plausible*.

THE OPPORTUNITY MINDSET

We present a new mindset for how entrepreneurs bring to life the most interesting and implausible opportunities. New models of opportunity are built on six insights very distinct from traditional approaches to strategy. Combined, these insights provide a fresh perspective of emerging trends in entrepreneurship and high-growth businesses. This book is structured around these insights to elaborate on how innovative entrepreneurs achieve the unexpected.

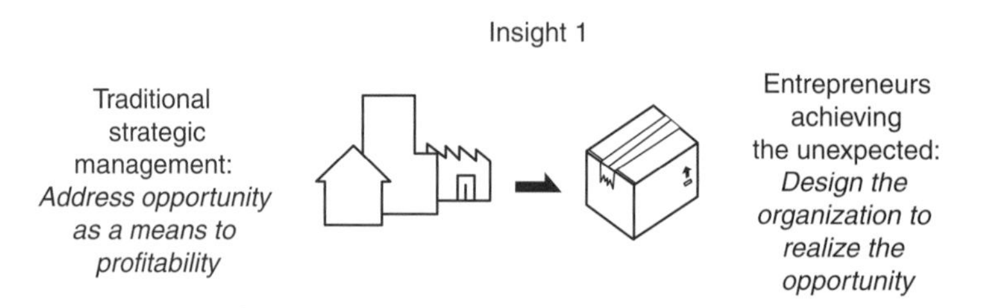

Profitability is the anchoring principle of competitive strategy. Ever-more sophisticated tools of risk-reward calculations attempt to provide analytical foundations for strategic decisions. In this framework, opportunity is the means to maximizing profitability. Managers select from a portfolio of opportunities to achieve profitability goals.

In the new world of opportunity and entrepreneurship, designing the organization is essential to reveal the nature of the opportunity itself. Profitability may provide no guidance to assessing the most interesting new opportunities. Monetizing interesting opportunities may be unknowable until the commercialization process is well under way. Entrepreneurs who focus on design as the anchoring element to realizing the opportunity open the horizon to the unexpected.

Traditional strategy treats opportunities as spheres – perfectly formed, and unchanging: diamonds waiting to be mined. The manager's challenge is clarifying the characteristics of the opportunity – to the extent that the opportunity is "well-characterized," the presumed accuracy of the mathematical tools provides the predictor of success. Just as managers engage in over-analysis, traditional entrepreneurship remains relatively focused on the benefits of planning and resource acquisition associated with "mining" these well-formed opportunities.

Insight 2

Traditional strategic management: *Perceive opportunities as perfectly formed ideas*

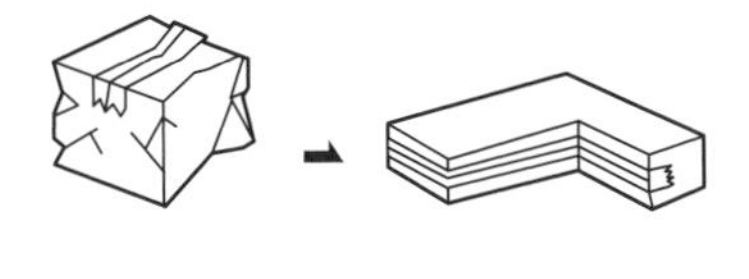

Entrepreneurs achieving the unexpected: *Appreciate the imperfections inherent to opportunities*

We think something very different is happening. Arbitrage opportunities in commodity or financial markets may be relatively featureless, but interesting opportunities accessible to the mass of entrepreneurs, especially in developing nations, will appear as imperfect shards – irregular and rough-edged. The inherent complexity of these opportunities, including geography, operations, risk, resources, and so on, will defy easy pigeonholing. Entrepreneurs will need to zoom in and out, rotate and reflect, dissect and invert opportunities to see different facets and levels of aggregation, without expecting to resolve a single, simple image.

Insight 3

Traditional strategic management: *Reconfigure resources to fit opportunities*

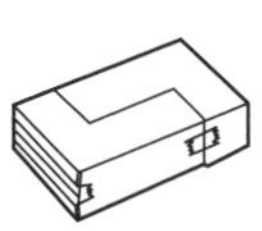

Entrepreneurs achieving the unexpected: *Remodel the organization for coherence*

The study of strategy as optimization has dominated organizational research for decades. And, for good reasons, the study of resources as a critical component of strategy formulation and implementation has heavily imprinted our understanding of competitive advantage. Firms with unique, valuable, inimitable resources win competitive battles. Exploiting new opportunities by leveraging existing resources makes sense in the context of fitting strategy to a competitive environment.

Our research suggests that this framework misses important structural elements that operate both within the firm and across firm boundaries. When firms address truly novel opportunities, coherence and plausibility, rather than fitness, play the critical role in whether the organization functions effectively. The growing challenge for new entrepreneurs and managers is not whether they accumulate the resources necessary to address a specific opportunity, but whether they can design coherent structures to address opportunities as they are revealed.

Insight 4

Traditional strategic management: *Build one business model for each opportunity*

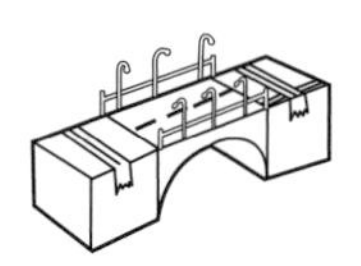

Entrepreneurs achieving the unexpected: *Build bridges to span opportunities*

Our interviews, experiences, surveys and analysis revealed that business models do not present an alternate strategic framework. Instead, business models are powerful levers to structure organizations around opportunities. We initially saw the wisdom of focusing on one business model for each opportunity. This idea has an obvious corollary: one business model per company, or at least per operating division.

But in an evolving world where entrepreneurs shape and are shaped by opportunities, this static conceptualization of opportunities, organizations, and business models focuses attention on exactly the wrong issue – the nature of fit. Fit presumes that for any given opportunity there is one optimized business model. The entrepreneur's role in designing the organization is not to fit a static business model to an unchanging opportunity. Her goal is to enable the organization to make the opportunity real, and ultimately capable of bridging from one opportunity to the next. A viable business model is a fine thing, to be sure, but few business models are defensible forever, and more likely than not, none are invulnerable. Building good

business models may be the realm of strategic analysis, but building bridges from one opportunity to the next is, for now, the province of the entrepreneur.

Seminal works on strategy argue that managers must adapt organizational structures to fit strategic design.[21] Although strategy formulation is both a manager-driven and emergent process, this heuristic encourages managers to exhaustively analyze business models to ensure fitness with specific characteristics of available opportunities.[22]

The evolving generation of entrepreneurs focuses instead on shaping opportunities and inspiring the narrative for change. Every entrepreneur perceives opportunities differently, and each brings a unique configuration of goals and preferences to the realization process. Business models are not simply structures to enable opportunities – they are levers for changing the underlying nature of the opportunities themselves. To inspire change and adopt new business models, entrepreneurs reshape the world around narratives that co-opt stakeholders in the pursuit of implausible opportunities. The words of Matthew Chambers, founder and CEO of Confederate Motorcycles, seem especially apt: "Fuel your inner man in revolt, be humble, reject comfort, and explore the myth."[23]

Of all the unexpected outcomes of our research, the most surprising and satisfying was that so many entrepreneurs approach the discovery process with a sense of wonder undiminished by the growing scope and complexity of organizational and global challenges. Though management practices emphasize profitability

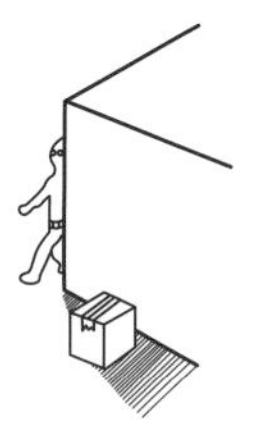

as the end justifying (mostly) reasonable means, these norms are being challenged from within and without, on a global basis, by a growing cadre of entrepreneurs born into a smaller, interconnected world.

They are the harbingers of broader change – and while they see themselves as agents of that change, they are also aware that their efforts both reveal and reshape the new world. They have fewer preconceptions about optimized business processes and outcomes; they see change not as a necessary evil for competitive survival but as a necessary good towards accomplishing entirely different results. In *Inventing Entrepreneurs*, we chronicled the journeys of scientists commercializing their own innovations. In this book, we confirmed that for the new generation of entrepreneurs, the journey that embraces unexpected opportunities is also the end in itself.

A BRIEF NOTE ON FORMAT AND CASES

We rely on prior studies and our own quantitative research to establish a broad, rigorous context for understanding opportunities, entrepreneurship, and business models. But we have chosen to avoid in-text citations of prior research and the documentation of well-established knowledge. Our intent is purely to facilitate a more coherent and fluid narrative. To be sure, we have included more "academese" than might be preferable for practitioners in some sections, and not enough rigor and specificity for our academic colleagues in others. Such flaws are probably unavoidable, but to the extent that what we have written is "interesting," we are pleased to be associated with the inherent imperfection of opportunity.

The case study companies are presented without judging either the capabilities or decisions of their management teams, or predictions of long-term prospects. As noted, the companies were not selected based on the anticipation of success or failure. We wanted to present a spectrum of industries and approaches, with a focus on relatively small, entrepreneurial organizations trying to accomplish provocative, engaging, and unusual goals. In addition, we needed complete, unfettered access to the executive team and the employee population, something not every firm will guarantee. Out of a larger population of companies, we selected 15 of interest, and were rewarded when 11 agreed to participate on our terms.

Because most, if not all of these firms will be unfamiliar to the casual reader, we have also applied the insights to more familiar, and, for the most part, more commercially successful, ventures. This is not to suggest that such firms succeeded because entrepreneurs or managers consciously applied the insights or analytics we propose. At the same time, we believe that the principles we discuss have been evolving in the social and structural spheres of the business community. Applying this lens to widely known examples may help other entrepreneurs and managers see and assess their own situations within a similar framework.

While many of the examples we discuss still frame organizational success in the context of profitability, some specifically and purposefully do not. The new world of opportunity extends beyond the boundaries of pure profit margins and revenue growth. Our ongoing research addresses these new, fascinating, and complex issues.

We hope this book stimulates new thinking about entrepreneurs, opportunity, innovation, and organizational action. We cannot hope to predict exactly which firms will succeed and which will fail, which innovations will be the drivers of sociological upheaval and renewal, or how the new world of entrepreneurship will change and manifest international politics and economics. We are certain, however, that business textbooks of the future will look back to this and the imminent decades as times of extraordinary uncertainty, change, and hope, driven by unprecedented entrepreneurial activity at all levels. In the meantime, we present this book to the new entrepreneurs, as well as to the rest of us who grasp at opportunities, however implausible they might seem.

NOTES

1 Barkema, H., Baum, J. A. C., and Mannix, E. 2002. Management challenges in a new time. *Academy of Management Journal*, 45: 916–930.

2 Reynolds, P. D. 2007. *Entrepreneurship in the United States*. New York: Springer:

3 The 2009 Global Entrepreneurship Monitor study found that while the new business ownership rate in the United States was relatively unchanged at 3.2 percent (consistent with Kauffmann), the rate in China has exploded to 11.8 percent, with 18.8 percent of the population now involved in early stage entrepreneurial activity. *2009 Global Entrepreneurship Monitor Executive Report*.

4 Khanna, T. 2007. *Billions of Entrepreneurs: How China and India are Reshaping Their Futures – and Yours*. Cambridge, MA: Harvard Business School Press.

5 Zooglobble interview with Cory Cullinan (2011) www.zooglobble.com/archives/2011/02/interview_cory_cullinan_doctor_noiz.html (accessed July 25, 2011).

6 "North State Symphony Receives Grant from McConnell Foundation to Support Classical Music for Children" (2011) www.csuchico.edu/news/archived-news/2011-spring/04–06–11-nss-receives-grant-from-mcconnell-foundation.shtml (accessed July 25, 2011).

7 In fact, the map is an approximation, because it only shows the Condor pools that report their presence to the University of Wisconsin; pools behind firewalls or under security restrictions don't appear.

8 Oviatt, B. and McDougall, P. 2005. Towards a theory of international new ventures. *Journal of International Business Studies*, 25(1): 45–64; Sapienza, H. J., Autio, E., George, G., and Zahra, S. 2006. A capabilities

perspective on the effects of early internationalization on firm survival and growth. *Academy of Management Review*, 31: 914–933.

9 Condor world map created by Douglas Thain, Todd Tannenbaum, and Miron Livny, based on the technique described in Thain, D., Tannenbaum, T., and Livny, M. 2006. How to measure a large open source distributed system. *Concurrency and Computation: Practice and Experience*, 8(15).

10 Aldrich, H. E. and Fiol, C. M. 1994. Fools rush in? The institutional context of industry creation. *Academy of Management Review*, 19(4): 645–670.

11 "30 million people online on Skype" (2011) http://blogs.skype.com/en/2011/03/30_million_people_online.html (accessed June 15, 2011).

12 Gompers, P. and Lerner, J. 2006. *The Venture Capital Cycle*. Cambridge, MA: MIT Press.

13 George, G. and Bock, A. J. 2008. *Inventing Entrepreneurs*. Saddleback, NJ: Prentice-Hall Pearson.

14 Aldrich, H. and Ruef, M. 2006. *Organizations Evolving*. New York: Sage.

15 Eckhardt, J. and Shane, S. 2003. Opportunities and entrepreneurship. *Journal of Management*, 29: 333–349.

16 Schumpeter, J. 1911. *Theory of Economic Development* (reprinted 2007). New Jersey: Transaction Publishers.

17 Kirzner, I. 1997. Entrepreneurial discovery and the competitive market process: an Austrian approach. *Journal of Economic Literature*, 35(1): 60–85.

18 Venkataraman, S. 1997. The distinctive domain of entrepreneurship research. *Advances in Entrepreneurship, Firm Emergence and Growth*, 3: 119–138; Sarasvathy, S. D. 2001. Causation and effectuation: toward a theoretical shift from economic inevitability to entrepreneurial contingency. *Academy of Management Review*, 26(2): 243–263.

19 Gulati, R. 2009. *Re-organize for Resilience*. Boston, MA: Harvard Business Press.

20 D'Aveni, R., Dagnino, B., and Smith, K. G. 2010. The age of temporary advantage. *Strategic Management Journal*, 31: 1371–1385; Barney, J., Ketchen, D., and Wright, M. 2011. The future of resource-based theory: revitalization or decline? *Journal of Management*, published online March 10, 2011, doi: 10.1177/0149206310391805.

21 Chandler, A. 1962. *Strategy and Structure: Chapters in the History of Industrial Enterprise*. Cambridge, MA: MIT Press.

22 Mullins, J. and Komisar, R. 2009. *Getting to Plan B: Building Better Business Models*. Cambridge, MA: Harvard Business School Press.

23 Matt Chambers speaking at the 2011 Gravity Free Design Conference, San Francisco, May 24–26, 2011 (speech text provided by Confederate Motorcycles).

1 Rethink organization design

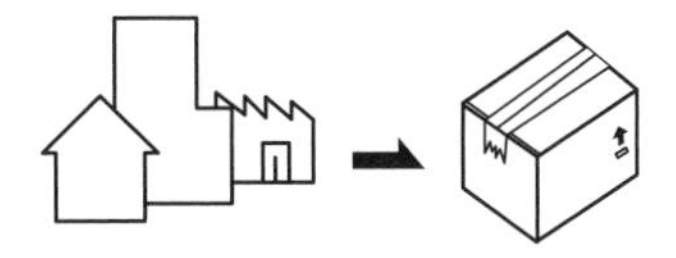

THE CENTER OF THE EMAIL UNIVERSE

There is almost nothing ordinary about Return Path (www.ReturnPath.net). Headquartered in the bustle and chaos of New York's Silicon Alley, most of the firm's operations are nestled in the serene foothills of the Rocky Mountains just west of Denver. This unlikely company's unorthodox organizational approach has brought an implausible opportunity to life. Return Path makes no apologies for its behavior, and no excuses about its plan to become the center of the email universe. Just the opposite in fact: the company vigorously touts itself as "the good guy" out to save the world from spam email. The company is still small, especially compared to global email services, interface providers, and the multinational ISPs that dominate the email world. Having doubled in size in 2010, Return Path is only now closing in on 300 employees, though the firm spans 11 offices across five continents. Technically, it still qualifies as a "small" business.

But Return Path's footprint is enormous. Led by visionary founders Matt Blumberg and George Bilbrey, Return Path's proprietary database on deliverability statistics broke the 1 *billion* email box barrier in 2009 and passed 2 billion in 2011. Quite an achievement for a "small" firm, considering that there are less than 3 billion valid email accounts worldwide! At the moment, Return Path can safely lay claim to more than 65 percent of the global email deliverability market. The rapid growth of Certified, the firm's email "whitelisting" system – yes, the opposite of blacklisting – amid the economic downturn of 2008–2010 suggests that adoption momentum is growing.

Tell someone that Return Path helps email marketers get their emails into one's inbox, and you might be rewarded with a frown of disapproval. It would be easy to think that Return Path is *contributing* to the spam problem. Overcoming this critical misperception is the key to the company's uniqueness and one aspect of its success. That's because email spam is Return Path's mortal enemy. One of the company's mottos is, in fact, "Keeping the world safe for email."

What does Return Path actually do? Perhaps the simplest way to appreciate the Certified program and its underlying email reputation database, Sender Score, is to compare it to FICO™ scoring. FICO™ is the credit rating system based on weightings of past credit activity developed by Bill Fair and Earl Isaac, the founders of Fair Isaac Corporation. FICO™ has become a global standard for financial institution lending decisions based on a sophisticated formula designed to represent whether an individual is a good or bad credit risk for a loan. Return Path's Sender Score performs a similar if spectacularly more complicated and extensive calculation – it compares data for email senders to industry aggregates and helps identify whether an email was sent from a reliable sender or a likely spammer.

To appreciate the difference in scale, consider that the 200 million consumers in the United States probably average about 1000 relevant transactions for FICO™ scoring purposes each year, though some are much more important than others. This suggests a computational requirement incorporating 200 billion transactions each year. It seems like a lot of data, and it is. But the email challenge is simply a different world. Recent estimates suggest that somewhere between 30 billion and 200 billion email spam messages are sent *every day.*[1]

Unlike traditional spam identification systems, which rely on identifying keywords (such as "Viagra" and, more recently, variations

FIGURE 1.1 Good email, bad email at Return Path
Image courtesy of Return Path, Inc.

like "V!agr0"), Return Path's anti-spam system is cumulative and more effective over time. Spammers can rapidly vary subject lines and body text, but emailbox addresses and IP addresses are much harder to fake, and over time patterns emerge for repeat spammers.

The entrepreneurial insight in Return Path's Certified program turns the traditional spam filtering model upside down. Return Path's founders realized that the long-term solution to the spam problem was tied into how we view email. Email filtering can look for "good" or "bad" email (Figure 1.1). Most spam filters catch bad emails by identifying questionable words or syntax in the email text. This effectively challenges spammers to develop more innovative ways to get around the filters. This has become especially troublesome as "bad email" has evolved from simple marketing spam to sophisticated phishing messages across email and IM delivery mechanisms.

Return Path identifies safe email by rewarding senders who conform to high standards for good email. This creates a "whitelist" of reliable email senders. Return Path works with ISPs, marketers, and the big email providers to create and enforce these standards. In other words, instead of playing the game by the spammers' rules, Return Path changed the rules of the game.

The Return Path business model is much more complex than this; it's one of the most sophisticated business models we've seen and we'll discuss it later on. But what's even more improbable is the venture-funded company's approach towards success. Blumberg and Bilbrey have applied novel and sometimes curious lessons to Return Path's growth and development. Rather than transplant structures and cultures from his prior successes, Blumberg decided that no matter what happened, he would encourage a positive, creative, supportive culture that put employees first. Incredibly, the company has a "What We Believe" video posted on its website that provides a public introduction to this vision; the Senior Vice President of People states, in no uncertain terms: "Return Path puts its customers second ... because we put our employees first." CEO Blumberg has repeatedly and publicly reinforced this perspective, and even added the fact that investors come third. In interview after interview with employees, we confirmed that this was the case. Is success as simple as world-leading technology and being nice to employees? The simple answer, of course, is no.

The Return Path elevator pitch, to solve the spam problem by establishing itself as the "good guy" at the center of the email universe, is both implausible and intriguing. As with many of the other firms we'll profile in this book, Return Path has already achieved improbable and unexpected results, and hopes to build on that success. So how does all of this fit together, and why does it work? What is Return Path's strategy – does it have a strategy? Why would a venture capitalist fund a company that explicitly provides so much extra support for its employees at obviously high costs? Any venture capitalist would realize that such policies create inherent strains in scaling growth. How does Return Path actually create value? How much of this value does it capture? What, finally, is its business model?

INSIGHT 1: RETHINK ORGANIZATION DESIGN TO REALIZE THE OPPORTUNITY

To achieve the unexpected requires approaching organizational design from a new direction. Traditional organizational design links firm structures to strategy: managers implement strategy by optimizing departmental, divisional, or matrix structures based on production and market characteristics. But in the emerging world of entrepreneurial action, where opportunities are almost unlimited in scale and scope, organization design has an entirely new purpose – realizing the opportunity itself. The design of organizational structures is directly linked to how the firm exploits a novel opportunity, long before strategic positioning against competitors in the market becomes relevant. Organizational design goes beyond formal hierarchy or even informal structures such as culture – it includes the structures that bound and direct transactions with partners and the systems that the firm uses to create value.

Matt Blumberg and the executive team at Return Path designed an organization that brought an opportunity into existence. They did this initially without directly considering product-market positioning or traditional aspects of long-term competitive advantage. They used an entirely different tool for identifying and developing key organizational characteristics and structures. This enabled an entirely new way of thinking about email marketing and spam. This powerful tool is poorly understood, in part because it has been described in far too many ways, using language that was designed for an entirely different purpose. That tool is the business model.

A DIFFERENT APPROACH TO BUSINESS MODELS

What is a business model?

Ask an entrepreneur to describe her firm's business model and you're likely to get a relatively quick, clear response. Entrepreneurs intuitively understand their firm's business model. But when pressed to explain details, or more importantly to explain what a business model is *in general*, even the most savvy and experienced executives may fumble.

Interestingly, academics and consulting gurus have not had much more success. The business model has been difficult to explain or measure in research for 20 years, dating back to the first mentions in research and general media. For some, business models are part and parcel of strategy. For others, it's just shorthand for the difference between revenues and costs. Rarely in social science research does one term become such a lightning rod for both acclaim and derision; rarely does so much research and practice activity fail to generate a clear understanding of a phenomenon that everyone seems to implicitly understand.

Business models have been a Pandora's box in academic research on organizations. Business models have been described as structures of transactions, sets of routinized activities, core value-creating propositions, processes for converting technologies into products, narratives of success, and much more. There are quite a number of ways to chart out business models in practice.[2] Just summarizing the descriptions of business models in research has become a challenge – searching for "business models" in academic research generates tens of thousands of publications! To provide a sense of the challenge, Table 1.1 provides a thematic summary of some of the most well-established literature on business models.

These academic definitions of business models are carefully worded and substantiated, but somehow seem to miss the point. They capture various facets of business models as theoretical objects of study, but do not seem to reflect how managers think about business models in practice. The concept of a business model emerged from business managers, not from deductive research. This is problematic, because much of the research on business models doesn't consider how managers actually use them. But even if the language used in these definitions seems a bit overdone, we can still get a sense of the core concept that emerges – *a business model provides a link between an opportunity and the organization that exploits it.*

For Confederate Motorcycles, the manufacturer of the $85,000 Wraith motorcycle, the opportunity is embodied in a relatively small group of super high net worth individuals, mostly men, wishing to express their individuality with tangible symbols of style, power, rebellion, and vitality. For Metalysis, a venture capital-

Table 1.1 *Thematic summary of business model research*

Theme	Summary	Representative definition	Sample publications[3]
Design	Agent-driven or emergent configuration of firm characteristics	"A business model is an architecture for product, service and information flows, including a description of the various business actors and their roles." (Timmers 1998)	Timmers 1998; Slywotzky 1999; Slywotzky and Wise 2003
Resources	Organizational structure co-determinant and co-evolving with firm's asset stock or core activity set	"Each business model has its own development logic which is coherent with the needed resources – customer and supplier relations, a set of competencies within the firm, a mode of financing its business, and a certain structure of shareholding." (Mangematin *et al.* 2003)	Winter and Szulanski 2001; Mangematin *et al.* 2003
Narrative	Subjective, descriptive, emergent story or logic of key drivers of organizational outcomes	"[Business models] are, at heart, stories – stories that explain how enterprises work." (Magretta 2002)	Magretta 2002

Table 1.1 (*cont.*)

Theme	Summary	Representative definition	Sample publications[3]
Innovation	Process configuration linked to evolution or application of firm technology	"The business model provides a coherent framework that takes technological characteristics and potentials as inputs and converts them through customers and markets into economic outputs." (Chesbrough and Rosenbloom 2002)	Chesbrough and Rosenbloom 2002
Transactive	Configuration of boundary-spanning transactions	"A business model depicts the content, structure, and governance of transactions designed so as to create value through the exploitation of business opportunities." (Amit and Zott 2001)	Amit and Zott 2001; Zott and Amit 2007, 2008
Opportunity	Enactment and implementation tied to an opportunity landscape	"[The business model] is a set of expectations about how the business will be successful in its environment." (Downing 2005)	Afuah 2000; Downing 2005; Markides 2008

funded UK firm developing unique metals processing technology, the opportunity is the limited availability of high-grade metals like titanium due to the resource and ecological costs of ore processing. For Praj, an Indian provider of biofuel and fermentation processing plants, the opportunity includes the growing need for both local biofuel processing facilities and higher-efficiency processing technologies.

Every business addresses some type of opportunity. Although entrepreneurship is often described as the discovery and exploitation of opportunities, businesses continue to address opportunities even when they are no longer acting entrepreneurially. At the venture formation stage, entrepreneurs may identify a previously hidden opportunity, or even help create the opportunity, perhaps by cultivating a previously nonexistent need. This can happen at a start-up like Confederate or at a fully established organization, as when Sony created the Walkman™. But opportunities continue to exist, though they may change, grow, and evolve. Some ultimately fade away, like the need for vacuum tubes following the development of transistor-based computer chips. When that happens, if the firm has not moved on to new opportunities, the firm will fail. But any firm still actively trading is exploiting an opportunity.

Equally important, every organization develops structures that coordinate the resources and activities to exploit that opportunity. The variety of opportunities in the world is only matched by the variety of characteristic structures used by organizations in pursuit of those opportunities. This is the goal of many research efforts – to categorize or compartmentalize extremely complex systems. It is effectively inevitable that some amount of data is lost in the process. But ordering and simplifying remains an incredibly valuable process towards better understanding and communicating complex systems. In the case of business models, a variety of such systems have been proposed, but none address how business managers talk about and use business models in the real world. To systematize this varied universe of opportunities and structures, we need to appreciate the language that managers use to describe and apply business models, and what their choice of words tells us about how they think about business models.

THE LANGUAGE OF BUSINESS MODELS

Rather than apply a theoretical framework to the question of business models, we simply went to entrepreneurs and asked them. We started with a small sample of venture capitalists and entrepreneurs. It was immediately apparent that none of the academic definitions for business models consistently reflected their descriptions. Here was a puzzle: given thousands of academic and practice articles written about business models, why didn't the views of management scholars and entrepreneurs match?

We decided to adopt a systematic and quantitative approach to understand the language of business models in practice. We administered surveys to more than 200 managers of British and Indian firms. We did not limit ourselves to small firms, because we believe that a broad definition for the business model should span all corporate sizes. We then assessed the words and language that managers use to describe business models. Box 1.1 provides a more detailed discussion of the research used to assess the language of business models in practice. It may be of methodological interest to organizational scholars and readers with rigorous interest in management research. Figure 1.2 provides a high-level summary of the results. The important conclusion is that business models are primarily about design and opportunities.

Although this analysis provides a useful starting point for understanding the language of business models, we further subcategorized the words used by managers to reveal the critical patterns and dimensions of business models in practice. Figure 1.3 shows how frequently those subcategories occurred in manager responses. The gray shape represents the frequency of concepts in manager descriptions based on a high-level analysis while the

BOX 1.1 A discourse analysis of business models in practice

We used discourse analysis to categorize and quantify the usage of word types to clarify and codify the language that managers use to describe business models. Our first concern was that the responses we obtained to the question "What is a business model?" were reflective of standard English usage. If the responses were out of line with spoken and written English, then the data would be non-representative of standard discourse and thought. Our analysis used 180 responses from entrepreneurs and senior executives who used 2417 words among them to define a business model.

Table 1.2 shows the occurrence of the 50 most frequent words from the responses. Word frequency and usage in the data sample are consistent with standard written English. For example, high-usage words that do not carry context-specific meaning, such as the word "and," occur with roughly the same frequency in the data sample as in standard written English. Words that carry meaning, such as the words "framework" or "process," represent a similar percent of the overall data sample as meaning-carrying words represent in standard written English.[4] Two small differences are worth noting: the highest-frequency words represent 23% of the sample but only 17% in standard English and the lowest-frequency words make up approximately 21% of the total sample compared to 47% in standard English. In other words, the sample is marginally less varied than written English, corresponding to the focused subject matter. Despite these minor distinctions, the sample provides an appropriate basis for assessing the language of business models.

We then analyzed the data to generate three types of results. First, we assessed each response to the question to determine which of the themes were represented. This provides a high-level assessment of the responses, which is a traditional qualitative tool. Second, we categorized every unique word in the sample into the six themes. This provides an assessment of the common language used by managers. Finally, we weighted that categorization by the frequency that unique

Table 1.2 *Fifty most frequent words describing "business model"*

Word	Frequency	Word	Frequency	Word	Frequency
the	151	for	18	your	12
a	117	organization	18	be	10
and	105	by	17	customers	10
of	100	on	17	framework	10
to	98	that	16	strategy	10
business	67	achieve	15	we	10
is	56	an	15	can	9
or	54	plan	15	one	9
in	45	you	15	are	8
which	41	market	14	do	8
it	31	revenue	14	make	8
value	28	services	14	products	8
company	27	with	14	service	8
how	21	customer	12	set	8
its	19	growth	12	structure	8
process	19	model	12	vision	8
way	19	product	12		

words occurred in the sample. This provides the most accurate reflection of usage by managers. Table 1.3 shows the results of the three levels of analysis, which reinforce the consistency of the responses.

outline shape shows the actual word frequency of those concepts. The strong similarity between the analyses shows that the categorization is consistent across levels of communication. The results reinforce the importance of design and opportunity, but show that the link between design and opportunity is organizational structure. A business model *designs organizational structures to enact an opportunity.*

Table 1.3 *Response, word occurrence, and word frequency by themes*

Theme	Responses	Words	Word frequency
Design	82	146	317
Resources	38	78	136
Narrative	14	48	62
Innovation	1	1	1
Transactions	57	100	209
Opportunity	59	107	264
Value	64	52	153
Total	315	532	1142

A more detailed description of this study and results is available in George, G. and Bock, A. 2011. The business model in practice and its implications for entrepreneurship research. *Entrepreneurship Theory & Practice*, 35(1): 83–111.

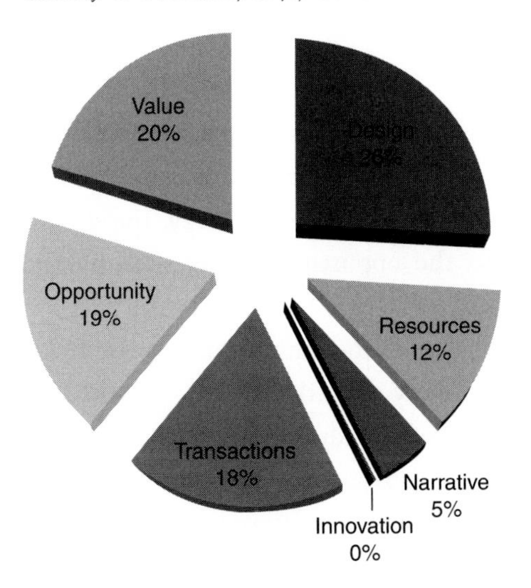

FIGURE 1.2 The language of business models in practice

In other words, *the language of business models is the language of opportunity, and the mechanism employed by managers to exploit an opportunity is organization design.*

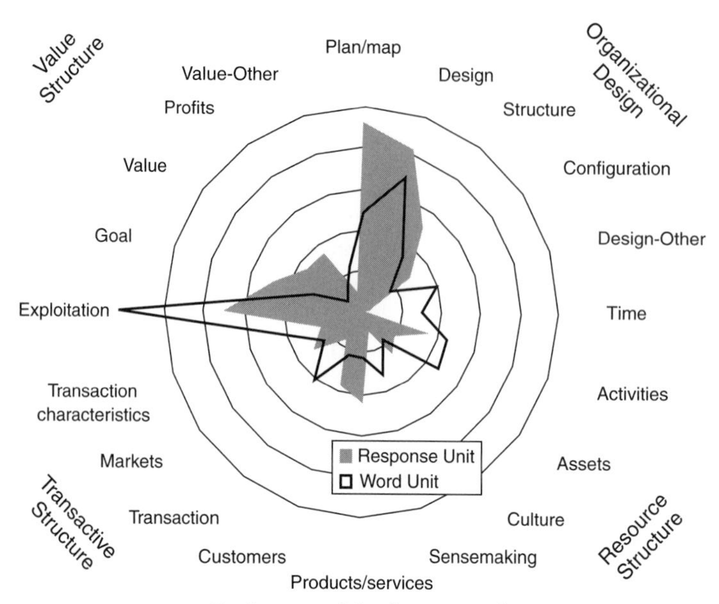

FIGURE 1.3 Business model subcategory elements

Although opportunities can come in many forms and types, we are focused on commercial opportunities – opportunities that support a for-profit business.[5] For managers, business models are a language for describing, usually in an abbreviated way, the design that links the organization to the opportunity identified by the entrepreneur.

Some opportunities are so well-defined that businesses addressing those opportunities tend to follow a similar design – sometimes referred to as the "dominant logic" of an industry. As an example, the buying and selling of residential properties in the United States was, for decades, dominated by the professional realty industry. Similarly, new automobiles in the United States are primarily sold through exclusive dealerships. In these fields, new opportunities were limited primarily to the selling capabilities of the agents, sometimes augmented by new features, especially for automobiles. But the advent of novel technologies can change this logic and bring new business models to the fore. The residential real estate industry in the US is under both pricing and channel pressure because the availability and searchability of property information on the Internet facilitates homeowners marketing their own homes. To a lesser extent automobile dealers have begun to adopt internet-based selling models for internet-savvy customers willing to conduct negotiations by email. The fact that common business models were in use by most companies reflected the stability in these industries. This isn't to say that truly innovative ideas *couldn't* have developed in the absence of technological change. But by most measures such changes weren't found and exploited in those industries until the mass adoption of the Internet.

The nature of entrepreneurship is the identification of opportunities. But identification is only the first step in the process. We're interested in process, not potential, and specifically the processes of organization-building and organization-changing. When Confederate Motorcycles founder and CEO Matt Chambers decided he wanted to resurrect classic ideals of American industrial design, he chose to do it by producing motorcycles unlike anything anyone had ever seen before (see Figure 1.4 for Confederate's B120 Wraith). Talk to Matt directly for more than five minutes, however, and you may begin to wonder if you're engaged in a conversation about business or comparative philosophy and human ethics:

> 60% of the world is growing very fast on its American System economic axis. The ironic oceanic landfill shall exponentially grow from such Taylor-made mindless efficiency unless we shift on the axis to the American Way: mindfully, effectively, artisinally, handcrafting high concept, high level design valuables, created to be permanent. The 'Less is More' result creates an economic growth model from the craftsmanship of fewer things which, ironically, result in far greater value in the lives of the clients we serve. Top line sales growth occurs, the craft wage explodes and all existence benefits.[6]

FIGURE 1.4 Confederate B120 Wraith
Image courtesy of Confederate Motorcycles, Inc.

For Matt, the opportunity was partly about selling high-end motorcycles to wealthy clients, and partly a statement of integrity, individual achievement, and the power of design. The business model of Confederate has not been focused on extracting every penny of profit from the sale of exotic street machines, though some might argue the firm would be more conventionally successful if it were. But the cycles never fail to elicit a visceral reaction. In that sense, the company has realized an opportunity, and the organizational design adopted by Confederate is an essential aspect of that outcome.

BUSINESS MODELS AS COGNITIVE MAPS

Approaching business models from within a cognitive framework resolves some problems while creating others. First, business models center on opportunity and thus are fundamentally entrepreneurial constructs. The important corollary to this conclusion is that a business model is not equivalent to firm strategy. Second, the managerial language of business models points towards understanding three dimensions of organization design: resource structure, transactive structure, and value structure. These will be helpful in providing a basis for comparing business model characteristics and even formalizing business models. Organizational structures associated with firm formation and change are closely linked to what managers believe about their business models and how managers describe business models. This is not too far from a description of the business model commonly, if uncertifiably, ascribed to Drucker: "the representation of how an organization makes money." It is very helpful to think about business models as cognitive artifacts, that is, simplifications or depictions of more complex sets of interactions and value-conveying resources. A firm's business model is the mapping and characterization of the key conceptual elements that describe how the firm operates.[7]

A business model is not a set of routines or resources. But entrepreneurs and managers use it to understand and communicate important information about routines and resources. In fact, business models are more subtle and complex than this. While managers may adjust a business plan to incorporate new information, they may also adjust their interpretation of new data to accommodate the dominant business plan at the organization. This may be a necessary aspect of understanding business models at entrepreneurial firms, where managers must accept inconsistent or even paradoxical intra-firm elements. Why? Because whether a business model works in a given context can't be determined *in advance*. The sense-making process is an essential component in how entrepreneurs resolve inconsistencies in information and thus both interpret and construct the context in which the venture develops.

BEYOND STRATEGY

Business models are not just a component of corporate strategy; neither is corporate strategy just a component of the business model. Strategy addresses firm competitiveness and performance in an industry context. In a cogent and well-known formulation, Professor Michael Porter argued that a significant amount of a firm's performance could be determined by the profitability of the entire industry and how well-positioned a specific firm was against competition within that industry. For example, pharmaceutical companies are, on the whole, more profitable than airlines. But even though the airline industry tends to generate low profits, Southwest Airlines consistently outperforms other airlines because it effectively employs a low-cost strategy that gives it distinct advantages over other airlines. It's not surprising, then, that it was Professor Porter who noted that business models were not strategy, if only because he had not seen a satisfactory definition for a business model.[8]

We agree – business models are not strategy.

Strategy is a dynamic set of initiatives, activities, and processes associated with the firm's position in a competitive context. A strategy may be reflexive, initiating change within the organization that impacts the emergence of strategic action. In contrast, the business model is a configuration of organizational elements and activity characteristics that comprises the firm's existence. A business model is inherently non-reflexive. Implementing a business model may lead to organizational change, but the business model itself is not a description of or a recipe for change. Business models are opportunity-centric, while strategy is competitor or environment-centric.[9] Understanding this distinction is critical, because general theories of entrepreneurial action have often framed entrepreneurship within a strategic context focused on wealth creation and optimization.[10]

What is the bottom line on business models and strategy? A business model is the key determinant in whether a company *may* operate successfully; strategy is the key determinant as to whether a firm is more or less successful than other companies in a given industry. Business models are not about competitive positioning. A business model is the organization's configurational enactment of a specific opportunity. Strategy is the process of optimizing the effectiveness of that configuration against the external environment, including the potential to change the configuration, alter the underlying opportunity, or seek out new opportunities.

The cognitive processes associated with opportunity identification and exploitation may not incorporate firm-level strategic thinking. Entrepreneurship has, in fact, often been defined as the study of why and how some individuals see opportunities, act to exploit those opportunities, and succeed.[11] But the decision to start a new venture, or redirect an existing company towards an entirely novel opportunity, creates a business model. Explicitly or implicitly, the business model applies organizational design to realize the opportunity, not establish corporate strategy. The business model is the core building block of the entrepreneurial process.

NO EXHAUSTIVE LIST

We started our research on business models by simply asking managers to explain their firm's business model to us. We got some surprising responses. The manager of the venture arm of a Global 500 manufacturer told us, in no uncertain terms, that there is only one business model – making and selling a product or service. Fascinatingly, the venture manager for a Global 1000 financial company located just miles away told us that business models were infinite – that every firm's business model was somehow unique. How could extremely smart executives in such similar functional roles present such different perspectives? Which one is wrong? Are they both wrong?

In fact, they're both right.

To understand why, we need to uncover the dimensions of a business model and the way those dimensions interact to shape how firms function, create value, and survive.

The business model designs organization structure to enact an opportunity.

We introduced this definition to show how business models link firms to opportunities. Our goal is to clarify those mechanisms and help you identify how those mechanisms function at your own firm. Before we do that, we need to explain why there isn't a definitive list of business models or business model building blocks to compare against your organization.

Lists like this do exist. In fact, in the e-business realm there's an entire book that develops the concept of atomic business model elements that can be combined into molecular business models.[12] Andersen Consulting developed an extensive list based on selling processes. These efforts present useful ways of thinking about one dimension of a business model – the transactive dimension. But they are not exhaustive, for two important reasons.

First, business models are not solely defined by the way a firm interacts with suppliers and customers. Certainly, there are similarities between firms that sell software and firms that sell razorblades, but claiming these are the same business model ignores dramatic differences in how these firms actually operate and create value. A business model must incorporate the underlying resources and capabilities of an organization, as well as something about how value is created. This means that any list of business models based on transaction types is incomplete.

Second, the endless cycles of industry, technology, and firm evolution result in new types of businesses with new business model elements that can't be predicted, much less categorized in advance. Auction firms and electronic transactions have been around for a while, but only the people who linked those transactive systems to online infrastructure thought to start companies like eBay. The same goes for Craigslist, an organization that combines online classified advertising with the apparently shocking business model characteristic of not maximizing profits.

New business models are born out of combinations of existing business models. The complexity of Return Path's business model requires understanding software services, the strange world of email spam, and the concept of critical mass in collective knowledge sharing across otherwise competitive internet service providers and telecommunications companies. Is it conceivable that a truly exhaustive list of all possible business models could be generated? Perhaps – but probably in the same way that, as Jorge Luis Borges described, a library of all possible books could be generated based on all possible combinations of letters and numbers. It would be finite, but so large as to be effectively uncountable.

As academics, we are sometimes criticized for trying to count angels dancing on pin heads. In this case, we acknowledge that we don't know, and may never know, the set of all "business model" types. But this doesn't prevent us from understanding key business model characteristics and dimensions.

BUSINESS MODEL DIMENSIONS

Even a nearly infinite library benefits from some kind of cataloging system. But the selection of the catalog terms or discriminating characteristics presents real challenges.

Although many companies use similar business models, the process of "lumping" together the functions that firms use to address opportunities may miss the most critical aspects of how business models actually work at firms. For example, Confederate's model of low-volume production, selling before manufacturing, and high-end products exists at other motorcycle companies and elsewhere. But without an understanding of the unique central role played by founder Matt Chambers, and the nature of the kind of value he wants to create, this analysis misses what makes Confederate's business model interesting in the first place. Equally important, it would entirely fail to explain the impact Confederate has had on the motorcycle industry distinct from profitability. Similarly, splitting Return Path into financial, operational, and marketing elements wouldn't identify the linkage between their cooperative relationship with ISPs that generates the whitelist database and the firm's revenue-generating relationships with email marketers.

Our approach, based on what managers and executives told us in interviews and surveys, is to understand the broader dimensions of business models, and how those dimensions interact to create something that is greater than the simple sum of the operational parts.

Business models have three dimensions. The first is the resource structure, which describes how a company organizes its assets and capabilities. The second dimension is the transactive structure – the architecture of the firm's interactions with partners, customers, and suppliers. The third dimension is the value structure – the design and purpose of the firm's system of realizing value from opportunities. These dimensions are inherent to any business model, even for *de novo* ventures. Firm formation establishes a resource structure, no matter how rudimentary; enactment of any opportunity establishes a transactive structure linking the firm and at least one external entity; firm viability requires a value structure that creates and captures some minimal value to replenish or augment the firm's resource base. This new language of opportunity points towards the entrepreneurial revolution that will dominate business in this century.

The first dimension: resource structure

Resource structure refers to the static architecture of the firm's organization, production technology, and core resources leveraged to serve customers.

Many business model analyses focus on the firm's product or production technology. This is consistent with contingency theory, which suggests that similar products and production technologies generate business models with similar characteristics. A significant majority of our survey participants mentioned product, production technology, or resource type in either the definition of a generic business model or a firm's specific business model. For example:

> *The process of employing capital and resources, people, process, and technology, to produce goods and services which will satisfy the needs of communities of customers thereby creating economic value for all the stakeholders involved.*

The business model "resource structure," however, is not simply the bundle of resources that the firm utilizes to create value. The resource structure describes the configuration of resources, capabilities, and activities associated with the firm's core business operations. The underlying elements of resource structure are, therefore, the general form of organizational structure, the nature of the firm's primary production systems, the structures that support the development and accumulation of critical value-bearing resources, as well as the implicit aspects of organizational structure, like culture, that coordinate activities. Each of these elements may be dissected into a variety of underlying organizational components, but the most interesting characteristics of resource structure can only be understood in the context of the target opportunity.

The decision to open an organics-focused co-op rather than a traditional convenience store is primarily a business model tradeoff, rather than a resource-based decision. A low-density architecture that engenders casual hierarchy, cooperative culture, and limited investment in infrastructure might be key components of the resource structure that co-evolves with the organization's activity bundles and value proposition. All of these may then feed into a strategic positioning of the business within the community market for groceries, perhaps as a high-price niche provider to a health-focused market segment. The resource structure describes the architecture in which the firm's strategic resources are embedded without necessarily determining or deriving from a strategic plan or decision.

It seems obvious that resource structure and resource strategy would co-evolve; so research on business model and strategy coevolution holds much potential. Similarly, the resource structure of early stage biotech and pharmaceutical companies may not be obviously linked with the firm's dynamic positioning within the industry and is more reflective of founder/entity opportunity enactment. In this case, resource structure and strategy intersect at the development of unique intellectual property that will determine whether a viable opportunity is successfully enacted. But some resource structures are more likely than others to enable the development process, regardless of the underlying value of the resources at stake or the specific strategic activities of the firm, such as network and partnership development.

The second dimension: transactive structure

Our investigation of business models traces its lineage to pioneering research by Rafael Amit and Christoph Zott, who studied the business models of internet companies at the height of the dot-com boom. They drew two important conclusions from their research. The first was that so-called dot-com companies were creating a different *kind* of value than other firms that couldn't be explained with strategic analysis. They suggested this was business model value, and that these firms were generating extraordinary value through new business models. A better explanation might be that this unexplained value was due to an investment bubble unsupported by fundamental indicators. The availability of capital from years of strong economic growth and the advent of the Internet likely fueled this bubble, overwhelming any clearly observable effects of business model innovation. In fact, many of the firms Amit and Zott studied between 1998 and 2001 had lost more than 99 percent of their market capitalization by 2002 or were gone altogether.

But Amit and Zott made a significant contribution to the research on business models, perhaps the most important to date, because they identified two key business model components. First, they argued that a business model was focused on exploiting an opportunity. Second, they showed that the set of transactions a firm developed with other entities (partners, customers, stakeholders) was a critical part of the firm's business model. We refer to this as a "transactive structure," which is the organizational configuration that determines key transactions with partners and stakeholders.

Our discourse analysis study reinforces the importance of transactive structure. This is well-aligned with rigorous studies on business models, but suggests the inclusion of the interactions between the firm and its key stakeholders – namely employees and shareholders. The transactive element of business models presents a macro-level architecture that can be directly linked to the firm's value creation outputs. This is particularly relevant for differentiating the variety of business models of firms utilizing novel information and communication technologies. Academic research provides a set of characteristics for transactive structures based on transaction cost economics[13] and business model-specific research;[14] the challenge lies in characterizing the structures, rather than the content of the transactions. Two of the firms from our pilot interviews develop and sell drug assay tools to organizations that perform high-throughput screening of drug targets. The underlying technologies are dramatically different, and the diseases for which the technologies are targeted are completely distinct, but the characteristics of the underlying transactions, and the organizational structures that configure those transactions, demonstrate significant similarities.

Even when cost structures and boundary-spanning activities differ, firms may display strong similarities in transactive structures, in part because mature markets and industries tend to reward standardization and conformance to dominant design. Most research on transactive structures has focused on industries in which significant value creation stems directly from the transactive role the firm performs, such as e-businesses. This has generated useful typologies of internet and IT firm types, as well as characteristics of successful firms, including efficiency and customer lock-in. But much less is known about the nature of transactive structures beyond e-business and technology industries more generally.

The third dimension: value structure

Value structure is the system of rules, expectations, and mechanisms that determine the firm's value creation and capture activities. A common element across executives' perceptions on business models is value, but business model value incorporates the structure of value creation and capture in the context of opportunity enactment. Value structure is the organizational system that defines, supports, and controls the processes through which value is created and then, ultimately, captured. Value structure serves as the facilitator between the nature of the underlying opportunity and the enactment of that opportunity via resource and transactive elements. It is the differentiating point of entrepreneurial co-creation that establishes the boundaries and enabling mechanisms for entrepreneurial action, mediating between the fundamental opportunity and the entrepreneur's perceptions of the opportunity landscape. As the firm acts to exploit the opportunity, the elements of value creation and capture likely adjust with the development of resources and boundary-spanning transactions. The value structure, however, may remain relatively constant, providing the high-level guidelines that link the entrepreneur's perception of available value to strategic decisions to maximize value creation and capture.

IT MATTERS

Venture capitalists, among many others, are interested in business models for two reasons. First, a firm's business model is a critical determinant of the firm's potential for survival and growth. Second, the way entrepreneurs and managers describe and use the business model provides important information about how they will operate the business. Much has been written about entrepreneurial success, often focusing on competitive advantage and the processes, mechanisms, knowledge and capabilities that help managers steer their companies to higher profitability and sector dominance. It's important not to confuse *viability* of the business with the *success* of the company. There are many reasons that a *viable* firm might not *succeed*: bad choices, poor execution, or misfortune, for example. *A viable business model is necessary, but not sufficient for a firm to survive.*

Equally important is the fact that *viable business models underpin organizations that succeed in non-traditional ways.* This is very different from the idea that firms that are not profitable do not have viable business models. We'll discuss this more towards the end of the book.

Our survey data affirms the importance of the business model and business model change. Managers understand that business models are the foundation of organizational survival. Changing a firm's business model can dramatically impact every aspect of firm operations and outcomes. But the majority of business model change activities we observed were *reactive,* and the modes of change and outcome were often highly predictable based on the perceived problem. This is important for managers to understand, because it suggests that in the case of business model change, managers and firms tend towards built-in, learned, or low-resistance responses. At best, this means that firms in a strategic or competitive group will tend to change business models in sync, so that little or no advantages will accrue to any one player. On the other hand, the one outlier, whether because of inherent creativity or simply random chance, may develop a novel business model and suddenly leap ahead of competition. These are the firms that achieve the unexpected or even change the very nature of the industry.

The economic crisis that started in 2008 presented more uncertainty and change than any other time since the Great Depression. During the 1957 recession a witty commentator noted that when General Motors sneezed the US economy caught a cold. In 2008 the US economy got the flu, and what was once unthinkable came true: General Motors declared bankruptcy; so did Chrysler. Lehman Brothers failed. Citibank, the world's largest financial institution in 2007, saw its share price drop from more $50/share to less than $1/share, shedding more than $125 *billion* in market capitalization, roughly equivalent to the entire economy of New Zealand. Global institutions that seemed fixed and permanent became suddenly shaken and at risk. National economies used to shrank; belts tightened worldwide.

In the midst of all this chaos, some firms seemed immune. In Fall 2009, Twitter raised private funds setting a firm valuation of $1 billion. Despite criticisms and problematic PR from a high-profile murder related to the use of the site, Craigslist revenues doubled to $100 million in 2009. Return Path doubled its emailbox coverage and employees while nearly doubling revenues in 2010. Of course, some firms have brilliant management teams that execute flawless strategies. But business models don't address relative performance against other firms; business models align the firm's structures against an opportunity in the broad industrial environment. A good business model isn't the difference between top-quartile performance and mediocre performance or bottom-quartile performance – a good business model represents viability. In other words, companies with good business models have the opportunity to succeed, while companies with bad business models simply fail.

EXPLORE, EXPLOIT, EXPLODE

In the world of competitive strategy, companies expend resources to explore or exploit opportunities. The balance between these two is critical. Too much exploration and firms are paralyzed; too much exploitation and firms miss emerging and more attractive opportunities.

This fundamental dichotomy is premised, however, on a world in which opportunities can be assessed strategically. In other words, a reasonable and rational process compares unfamiliar situations or novel markets against the familiar backdrop of experience or strategic frameworks to develop a cohesive plan. Everything is assessed in context – the context of the assets already in place at the firm, the capabilities and skills developed to get the firm to where it is, and the existing rules and boundaries for the broader industry.

The democratization of opportunity access and information has two implications for this well-established balance. The first implication is that regardless of how firms adjust the balance of exploration and exploitation, they will not be able to assess new opportunities in the context of existing operations or familiar circumstances. Why? Because some opportunities will be so novel that there will be no relevant context.

Second, firms will sometimes need to use a completely different process to address novel opportunities – they will need to *explode* them.

As much as it has created untold amounts of information value, Google has become increasingly adept at destroying value in adjacent industries. TomTom and Garmin lost nearly $2 billion in combined market value on the day Google announced it would make Google Maps available for free on phones running Android.

But one of the greatest opportunity exploders of the past decade has been Craigslist. US newspapers had been experiencing a gradual decline, though profitability had been held somewhat stable via consolidation. The decline in classified advertising is nothing short of catastrophic – peak revenues of $20 billion in 2000 dropped by 70 percent to $6 billion in 2009.[15] Craigslist is generally credited with a significant role in this collapse, as the leading online classified ad site in America, despite only generating about $80 million in revenues in 2008. After all, the vast majority of the 50 million ads posted to Craigslist *each month* are free. Craigslist could have implemented a business model of charging something, *anything*, for posting or responding to ads – even a penny. But rather than explore or exploit the market, Craigslist simply blew it up.

And it worked. Craigslist leads the market, though new entrants Oodle and Kijiji are growing rapidly. In 2008, the *Business Insider* calculated Craigslist's market value at roughly $5 billion. By comparison, the *New York Times* was worth roughly $1.6 billion in early 2010. Lee Enterprises, which owns newspapers in more than 40 small and mid-sized markets, has a market cap of roughly $150 *million*. But Craigslist wasn't out to maliciously run newspapers into the ground – the company sees itself as the good guy: the great equalizer in linking buyers and sellers for a nearly zero variable cost. *Business Insider* argues that Craigslist isn't run like a normal organization: "It's run like a non-profit. Craig Newmark and co. don't give a damn about generating revenue or profit, and more power to them."[16] The question isn't whether value creation took place – the question is whether the opportunity for classified ads looks anything like it did ten years ago. Perhaps Oodle, Kijiji, or some other firm will find a way to remonetize classified advertising, but at the moment, the Craigslist business model is working.

True opportunity exploders are rare. Exploders require a fully disruptive business model combined with an executive team and shareholders who aren't necessarily attempting to maximize near-term or even mid-term gain. These are distinct from new entrants with disruptive technologies, as epitomized in Christensen's *The Innovator's Dilemma*. Disruptive technologies use reconfigurations of improved existing technologies to target novel market segments that are still in the formation stage. In doing so, new entrants benefit from inertia at incumbents who cannot effectively allocate resources to nascent target segments, because their internal incentive and reward structures are generally designed to encourage success with large current customers. This is, however, the first hint at the challenge posed by disruptive business models targeting novel opportunities. Some disruptive innovations outdate existing technologies; some make incumbent capabilities obsolete; and some do both.

One of the companies in our study has a potentially disruptive technology. Metalysis, a UK-based technology firm spun out of Cambridge University, has developed what may be the first truly novel advance in metal ore processing since the advent of the Kroll cycle more than 50 years ago. The company's FFC Process™ could enable low-cost processing of high-value ores like titanium and tantalum. At the same time, FFC could also dramatically reduce negative environmental impact. Success for Metalysis means making high-value ores available at a lower cost for more applications; but this won't wipe out the major players in the industry. Metalysis can't blow up the metals market because there is too much embedded infrastructure and there are limitations on logistics of ore mining and transportation. Metalysis couldn't scale up facilities fast enough to bring costs down by two orders of magnitude. As exciting as the Metalysis opportunity is, it's not going to explode the market for titanium. Exploding a market requires a very unique business model in a very unique industrial context.

On the other hand, in a very real sense iTunes blew up the music distribution market. The price of music hasn't really changed dramatically; the cost of an average album has only fallen by a few dollars. But the implications of re-enabling the purchase of single songs, a capability that hasn't been around since long-play records replaced 45s, are much more far-reaching. iTunes changed the distribution of power in the recording industry. It has enabled some bands, like Radiohead, to simply drop out of the music studio system altogether – it released *In Rainbows* online and allowed customers to pay whatever they wanted rather than demanding a specific price. Whereas commercial music applications were once dependent on the expertise of the music publishing catalog experts, music supervisors at film, TV, record labels, and advertising agencies now have a wide set of sources for musical content, including unsigned artists. This disruption facilitated the success of Broadjam, one of the only online music companies founded during the dot-com boom to survive the industry shake-out, lawsuits, and realignment. Broadjam's story of forging an entirely new business model around unsigned musicians is told in the next chapter.

New entrants, and sometimes even incumbents in the new opportunity landscapes of the twenty-first century, companies like Craigslist, Google, and perhaps Return Path, present much more difficult profiles for competitors to assess. This is precisely because they are not just jockeying for competitive position: they are attacking opportunities for which they have unique access. Certainly Google sees competition from Yahoo! and Bing, but only Google is leveraging its massive advantage in total data accumulation to target unformed opportunities. Google has forward integrated into the application space with Google Docs, and backward integrated by introducing Chrome and Android as operating systems. Truth be told, management scholars have no technical term to describe Google's forays into nearly every aspect of the information and communications technology space. Apple may be the most valuable technology company, but it is hard to argue with the suggestion that Google has come to dominate the global sphere of information access and utilization. But Google Earth, Google Voice, mobile translation services, and Google Scholar all represent forays into unknown opportunity spaces that only Google has the unique data resources to leverage. Just as we complete this manuscript, Google has launched Google+ to compete directly with Facebook, having failed with its first social network incarnation, Buzz. Most assessments of webuse put Facebook and Google as the #1 and #2 websites in terms of utilization. How this particular battle will play out remains to be seen, especially as the underlying business models of the firms appear to be converging.

Entrenched firms face an entirely novel challenge: fighting a market entrant who chooses to compete based on a business model so bizarre or unlikely that it engages an entirely new set of rules. What happens when the entrant refuses to treat extant firms in the industry like competitors? Established firms face a second, even more problematic challenge: these entrants may simply explode the entire opportunity set rather than divide it up with current competitors. This goes far beyond competence-destroying[17] or disruptive innovation[18]. Whereas those innovations build new value creation propositions on the ruins of the old, some types of business model innovation obviate the value proposition or the exploitation mechanism altogether.

A quick thought experiment shows why this is so non-intuitive. Let's assume Craigslist had chosen to charge 10 cents for posting

Table 1.4 *Three stages in achieving unexpected outcomes*

	Explore	Exploit	Explode
Resource structure	Identifying novel aspects or uses for resources	Developing the capabilities to use and apply novel resources	Scaling resources while controlling access
Transactive structure	Identifying unfamiliar or newly enabled transaction methods	Convincing participants and partners to try novel transactions	Leveraging transaction capacity via open systems or network effects
Value structure	Identifying previously unavailable or unneeded value	Legitimizing novel forms of value	Creating self-feeding value creation cycles where value automatically creates more value

each advert. Let's further assume that the logistics of making this payment, presumably through PayPal or other electronic debiting, would have deterred half of all postings (this is likely conservative, since it seems reasonable to expect that most posts on Craigslist are from repeat users). This calculation suggests that Craigslist could be generating $30 million, or nearly 40 percent more revenues per year. It may seem incredible that a for-profit firm would be willing or able to leave money on the table, but we're already seeing that the classic rules of market entry, product pricing, and industry dominance simply won't always be subject to the dictates of "common sense."

How then do business models relate to how firms explore, exploit, or explode opportunities? Table 1.4 summarizes how the firm's resource, transactive, and value structures must be applied to exploring, exploiting, and exploding opportunities.

Not common sense ...

Clearly, designing the organization to address opportunities is not a simple challenge.

In fact, one of the first preconceptions that entrepreneurs relinquish to achieve the unexpected is the adage that common sense should drive all decisions. Common sense is a powerful tool for assessing the known and the familiar. Common sense leads to successful decisions that combine familiar elements in unfamiliar ways or unfamiliar elements in familiar patterns. But common sense has limitations. These limitations have been documented in scientific and popular literature, perhaps most famously in Malcolm Gladwell's book on intuition, *Blink*. Gladwell's argument is that subconscious intelligence often works dramatically faster than conscious logic, arriving at useful and correct choices at the gut level long before rational thought. These subconscious processes rely on stored information not easily accessible at the conscious level, such as learned visual cues for body language or other natural-world phenomena. A broader interpretation of these kinds of capabilities, usually defined as "intuition," plays an important role in understanding human behavior and decision-making.

But common sense and even intuition require context, and the developing economic environment of global opportunities will, at least for a while, be contextually ambiguous at best, and likely totally unfamiliar for most businesspeople, especially in developed nations and at well-resourced corporations. The time has come to prepare to put aside both complex and simple rules for how to run a business, and open ourselves up to thinking about how to design a business that functions and adapts very differently. Rather than common sense, managers of the most interesting entrepreneurial firms need design sense.

... Design sense

What is design sense? Design sense is a different approach towards thinking about organizational structures.

First, organizations incorporate two interconnected human systems: formal and informal structure. Formal structure usually refers to the hierarchy of the organization, while informal structure describes the intangible aspects of how the organization operates – characteristics such as culture, creativity, authority, and power. The critical aspect of organizational design that trumps "common sense" is that organizations, and the people within those organizations, can't be treated as perfectly rational and efficient mechanisms.

So common sense tells us things like:

- People look out for their best interests
- organizations are only as strong as the weakest link
- when in doubt, don't!
- take care of the customer and the rest will take care of itself.

Design sense tells us very different things:

- Organizational design will influence the priorities of employees
- Radical design changes can result from apparently small changes in resources
- Design is pervasive in the narrative used to explain how an organization works

Design sense approaches organizational management, and *opportunity management*, with a very different perspective. Business models that incorporate design sense tend to have explicit links between the underlying priorities of the organization and the structures and systems used by management to create value. Throughout this book, we'll come back to these key aspects of business models. We'll talk about why entrepreneurs can successfully address imperfect opportunities to build coherent organizations. We'll show how firms address new and changing opportunities by building bridges. And we'll consider the role of narrative in organizational design and management.

For now, the importance of organizational design can be summarized with three critical lessons:

1) Organizational design should reflect opportunity
2) organizational design can be fitted to opportunities or used to shape the opportunity
3) organizational design determines how the firm identifies and addresses opportunities.

DESIGNING THE UNEXPECTED AT RETURN PATH

Many aspects of Return Path's success could be attributed to the classic elements of successful entrepreneurial development. After all, co-founders Matt Blumberg, George Bilbrey, and Jack Sinclair all came from prior successful ventures in a technologically disruptive field. They've worked hard to bring together strong executive leadership including smart and committed venture capital. The company has established a culture of transparency and flexibility to deal with a complex environment.

Why, then, at a time when email growth is expected to double from 250 billion messages per day in 2009 to 500 billion messages per day by 2013, is a relatively small, relatively unknown firm leading the quest to tame the spam problem? Why, unlike most of the rest of the industry, is it using a "positive" role model framework rather than the "negative" spam keyword identifying system that is the dominant technology? And why is it happening at *this* firm, rather than any one of the literally hundreds of firms dedicated to dealing with spam?

In Return Path, the answer lies in a managerial perspective that *presumes* that understanding today's email-related problem is only a necessary capability for identifying the email challenges yet to come. For the Return Path executive team, the problem isn't optimizing the organization to help clients address spam – the problem is designing an organization that will be able to change the nature of the email spam war. Over the past ten years, what began as guerrilla-style spamming has turned into a full-fledged arms race in which, realistically, spammers have most of the advantages. Spammers can be mobile and flexible and have a low cost of attack. The individuals and corporations fighting spam are, for the most part, immobile, slow to uptake new infrastructure, and have no counterattack options.

Return Path hadn't even intended, originally, to fight this particular war. The company had morphed from a relatively simple email address forwarding company to a provider of email marketing services, when the nature of the war emerged to Blumberg and other key managers. They realized that seeing the landscape of opportunity associated with email spam required a completely different perspective – rather than winnow the spam *out*, they wanted to sift the good email *in*.

Return Path saw *through* the opportunity of the spam problem and turned it upside down. The problem isn't spam – it's the challenge of identifying reliable email senders. The goal of the company isn't to identify and eliminate spam, the goal is to make spam irrelevant by escorting *safe* mail into the inbox.

As the new organization emerged from the original business via a series of acquisitions and divestitures, Blumberg continued to reinforce the company's core values of employee empowerment and rapid adaptation. He wanted an organization that could constantly search for related opportunities without regard to the firm's core resources or expectations. The company took a seemingly extraordinary gamble – it partnered with email recipient systems, primarily large ISPs, to begin building a global database of email inbox data to identify the broadest possible trends of email deliverability. This required a significant investment in relationship building and trading of aggregated data, and would only work if the company achieved a critical mass of inbox data, without ever knowing what that critical mass level would be.

Blumberg and his team designed the organization to incorporate two distinct groups with characteristics that would be inherently difficult to motivate, reward, and control consistently. The "Receiver" team needed to build long-term relationships with global ISPs for the purpose of freely sharing data. The team would generate no revenues, incur significant costs, and attempt to encourage cooperation among large, viciously competitive organizations with high infrastructure and razor-thin margins. In the process, the Receiver team needed to create the world's largest aggregated database of email inbox data to enable Return Path's vision of "making the world safe for email." That database would be the foundation for establishing global standards for good email practice based on standards-based protocols and systems. The Receiver team could then compare aggregate anonymized statistics against the specific data provided by ISPs. This would help clarify how good and bad email behaved across a wide spectrum of end-user contexts, as well as identify high-reliability emailers who met the incredibly rigorous qualifications required to join the Certification program. Once all this was in place, the "Sender" team had to operate as a sophisticated email deliverability tools and consulting service. The Receiver side is effectively non-profit, so Return Path generates revenue selling high-margin software and advice to email marketers. It does so by encouraging them to aspire to the highest level of integrity. By conforming to a rigorous set of standards, those firms increase the probability of their marketing emails reaching their customers.

How did they manage the incredible internal conflicts associated with this structure? We're going to talk more about "frustrated systems" and "coherence" in Chapter 3, but the simple explanation is that the executive team realized that there would be too many differences in hierarchy, position, and role to either simply sweep under the rug or completely resolve. Once again, they saw *through* the challenge and simply came out the other side – they made the entire organization transparent, acknowledged the conflicts, and resolved them by elevating the welfare of the employees above the specific details of the revenue model itself. In other words, when in doubt, *do*.

Return Path demands extraordinary efforts from employees, but first it only hires employees after an extraordinary interview process. Multiple visits, shadowing a current employee, and a meeting with two top executives, including Blumberg himself, for potential hires at every level of the organization are the rule, not the exception. The culture at the organization constantly reminds employees of their importance: employees are actively encouraged, if not required, to identify additional talents and skills they'd like to develop, and then relevant training is provided. It's arguably no accident that Return Path's Senior Vice President of People is a former Regional Director of People Innovation for Kimpton Hotels – the hotel chain long-recognized for a unique combination of customer service and quirkiness. And, yes, Return Path had two full-time human resource managers overseen by a Vice President of People when there were only 100 employees.

Does Return Path walk the walk? Amazingly, it does. Whereas many early stage ventures find they have to tighten up loose human resource policies as the organization grows and matures, Return Path once again chooses to see entirely different perspectives. A few years ago, the company grew past 150 employees, the "Dunbar number" for how many people an individual can realistically interact with effectively. Traditional strategic thinking suggests that more structure is required to manage and monitor employee behavior. Return Path executives, on the other hand, decided that the firm's vacation policy didn't reciprocate employee commitment to the organization. So they eliminated it. Return Path employees take holiday time when they need it.

Need more proof? In 2010, Return Path executives faced a difficult decision. Because their business model integrates two distinct business systems – email receivers and senders – the interests and priorities of their stakeholders may sometimes appear to be at odds. Their receiver base, including ISP partners and, by extension, email users, was frustrated with third-party marketing email. In most cases this email represents a relatively small percent of total volume, and in most cases is both legal and ethically derived from email recipient actions. But to most consumers, it looks like spam, and it generates a high volume of complaints. But many of Return Path's customers either used third-party marketing or had relationships with third-party marketers. An internal assessment suggested that a change in policy could affect as much as 5 percent of the firm's revenue stream. Losing that revenue would wipe out the firm's profitability, just as the executive team envisioned significant reinvestment in operations.

What did Return Path do? It asked its employees to comment on the situation. It laid out the details, and the potential implications for the business. And the overwhelming response from every level of the organization was that Return Path employees were the "good guys" in the fight against spam, and anything less than the high road simply wasn't acceptable. To be sure, they notified their customers in advance, had extensive conversations with key accounts, and then buckled down to prepare for the worst. On March 24, 2010, the company decertified all third-party marketing email.[19] Blumberg recounted the moment:

> We flipped the switch on it ... we all sort of braced ourselves for, all right, we could lose $1.5 million in revenue tomorrow. And we lost none. There was some noise, there was absolutely some noise from this particular segment of the client base. But we didn't lose anyone. We were obviously happy we didn't lose anyone, but we were even happier for three reasons. One, was we had taken a risk and it had paid off, and that always feels good. Two, we had done something that we actually felt at the end of

> the day was fundamentally good for the email system. The third thing to feel good about was, our mission statement is about setting standards for the industry, and the more we do that the more it's clear we are a relevant player for the industry.

It is this element of Return Path's business model that holds together the conflicting elements within the product development structures, which would undoubtedly be a huge source of tension and inter-organizational stress otherwise. Return Path operates one of the most complex and intricately balanced business models we've ever seen – value, transaction, and resource structures operating at a finely tuned equilibrium.

Will it work? Right now Return Path represents less than 1 percent of the sales of the anti-spam industry. But Return Path has something most anti-spam businesses can't address – a business model that becomes *more* effective offering solutions that will be *more* cost-efficient as the spam problem increases. Traditional spam solutions will require ever-increasing storage and bandwidth to filter out ever more sophisticated spam messaging tactics. Think about your own anti-virus software options – the more sophisticated they get, the more they slow down your computer. Centralized and cloud-based computing has already begun to offer temporary efficiencies in combating these problems, but it seems equally likely that a standards-based system that relies on sender reliability data, rather than content scanning, will ultimately be the most cost-effective and stable solution. If so, nearly every email in the world could be touched by the far-sighted company in Colorado with the implausible plan of becoming the center of the email universe by putting the welfare of its employees before that of its customers.

NOTES

1 Symantec Corporation website: www.symantec.com/connect/blogs/spam-rustock-lethic-and-xarvester-disappears-over-holiday-season (accessed January 22, 2011).

2 Osterwalder, A. and Pigneur, Y. 2010. *Business Model Generation: A Handbook for Visionaries, Game Changers, and Challengers*. New York: John Wiley.

3 Timmers, P. 1998. Business models for electronic markets. *Electronic Markets*, 8(2): 3; Slywotzky, A. 1999. Creating your next business model. *Leader to Leader* (11): 35–40; Slywotzky, A. and Wise, R. 2003. *How to Grow When Markets Don't*. New York: TimeWarner Books; Winter, S. G. and Szulanski, G. 2001. Replication as strategy. *Organization Science*, 12(6): 730–743; Mangematin, V., Lemarie, S., Boissin, J. P. *et al.* 2003. Development of SMEs and heterogeneity of trajectories: The case of biotechnology in France. *Research Policy*, 32(4): 621–638; Magretta, J. 2002. Why business models matter. *Harvard Business Review*, 80(5): 86–92; Chesbrough, H. and Rosenbloom, R. S. (2002). The role of the business model in capturing value from innovation: Evidence from Xerox Corporation's technology spin-off companies. *Industrial and Corporate Change*, 11(3): 529–555; Amit, R. and Zott, C. 2001. Value creation in e-business. *Strategic Management Journal*, 22(6–7): 493–520; Zott, C. and Amit, R. 2007. Business model design and the performance of entrepreneurial firms. *Organization Science*, 18(2): 181–199; Zott, C. and Amit, R. 2008. The fit between product market strategy and business model: Implications for firm performance. *Strategic Management Journal*, 29(1): 1–26; Afuah, A. 2000. *Business Models: A Strategic Management Approach*. New York: McGraw-Hill; Downing, S. 2005. The social construction of entrepreneurship: Narrative and dramatic processes in the coproduction of organizations and identities. *Entrepreneurship Theory and Practice*, 29(2): 185–204; Markides, C. 2008. *Game-changing Strategies: How to Create New Market Space in Established Industries by Breaking the Rules*. New York: Jossey-Bass.
Source: George, G. and Bock, A. 2011. The business model in practice and its implications for entrepreneurship research. *Entrepreneurship Theory & Practice*, 35(1): 83–111.

4 Leech, G., Rayson, P., and Wilson, A. 2001. *Word Frequencies in Written and Spoken English: Based on the British National Corpus*. London: Longman.

5 Of course, every organization has a business model, but our data is based on for-profit organizations. We believe that this analysis extends

naturally to other organization types, including not-for-profits, but that is an exercise for another day.

6 Matt Chambers speaking at the 2011 Gravity Free Design Conference, San Francisco, May 24–26, 2011 (speech text provided by Confederate Motorcycles).

7 Baden-Fuller, C. and Morgan, M. S. 2010. Business models as models. *Long Range Planning*, 43(2–3): 156–171.

8 Porter, M. E. 2001. Strategy and the Internet. *Harvard Business Review*, 79(3): 62–78.

9 Nag, R., Hambrick, D. C., and Chen, M.-J. 2007. What is strategic management, really? Inductive derivation of a consensus definition of the field. *Strategic Management Journal*, 28(9): 935–955.

10 Hitt, M. A., Ireland, R. D., Camp, M., and Sexton, D. L. 2001. Strategic entrepreneurship: entrepreneurial strategies for wealth creation. *Strategic Management Journal*, 22(6–7): 479–491.

11 Baum, J. R., Frese, M., and Baron, R. A. (eds) 2006. *The Psychology of Entrepreneurship*. Bowling Green, OH: Society for Industrial and Organizational Psychology.

12 Weill, P. and Vitale, M. 2001. *Place to Space: Migrating to E-business Models*. Cambridge, MA: Harvard Business Review Press.

13 Williamson, O. E. 1991. Comparative economic organization: the analysis of discrete structural alternatives. *Administrative Science Quarterly*, 36(2): 269–296.

14 Amit, R. and Zott, C. 2001. Value creation in e-business. *Strategic Management Journal*, 22(6–7): 493–520; Zott, C. and Amit, R. 2007. Business model design and the performance of entrepreneurial firms. *Organization Science*, 18(2): 181–199.

15 Source: Business Analysis and Research, Newspaper Association of America.

16 Source: "Craigslist valuation: $80 million in 2008 revenue, worth $5 billion," Henry Blodget, *Business Insider*, April 3, 2008. http://articles.businessinsider.com/2008-04-03/tech/29995803_1_craigslist-users-craiglist-craig-newmark (accessed September 19, 2011).

17 Tushman, M. L. and Anderson, P. 1986. Technological discontinuities and organizational environments. *Administrative Science Quarterly*, 31(3): 439–465.

18 Christensen, C. 1997. *The Innovator's Dilemma*. Boston, MA: Harvard Business Press.

19 "Changes to Return Path Certification Standards," Return Path blog, March 24, 2011 www.returnpath.net/blog/intheknow/2010/03/changes-to-return-path-certification-standards/ (accessed July 25, 2011).

2 Appreciate imperfect opportunities

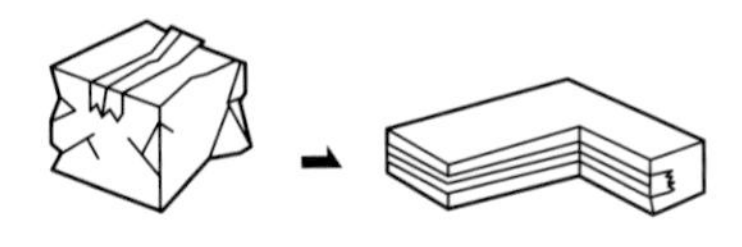

A DIFFERENT DRUMBEAT AT BROADJAM

At the turn of the millennium, the global music business, one of the most glamorous and tightly controlled industries in the world, entered a period of tumultuous, unprecedented, and unpredictable change. Prior evolutions of media format – vinyl, tape, compact disc – hotly debated but ultimately embraced by the industry, simply didn't prepare music industry executives for the advent of the Internet and the revolution in music distribution. The rapid adoption of web technologies and file-sharing protocols fostered a rainbow of online music sites: artist-specific streaming sites, peer-to-peer networks like Morpheus, centrally hosted file-sharing systems like Napster, and music distribution channels like mp3.com. Sophisticated royalty-sharing agreements that had evolved over nearly a hundred years of relatively slow industry change were suddenly outdated or even irrelevant when content could be traded instantaneously and almost untraceably. The major record labels and publishing companies, apparently unprepared by the massive groundswell of file-sharing and piracy, were slow to react. That lack of clear response lent an aura of inevitability to predictions of a complete industry restructuring.

In the midst of this chaos, Roy Elkins founded Broadjam Incorporated (www.Broadjam.com), an online music sourcing and distribution site targeting the vast population of music hobbyists and independent musicians. These individuals range from highly trained musicians to self-taught hobbyists and a new generation of electronic music composers, all enabled by powerful and inexpensive music production technologies associated with home computers. But the vast majority of musicians don't sign contracts with a recording label or a publishing company. And for decades, the link between unsigned musicians and a recording contract was the narrowest of filters – a very small group of talent scouts at key industry firms.

While the major industry players entered a protracted war over the rights and returns for the big name artists and major labels, Elkins had a different vision in mind: redirecting the same internet systems to benefit independent musicians looking to make a living, or perhaps even get that critical big break into the industry. Broadjam could leverage Elkins' music industry connections, create a platform for distribution, and establish processes for identifying and improving talent. Elkins' original vision was to collect independent musicians' compositions, catalog the music, and then blast it to online distribution entities such as mp3.com where it would be monetized via advertising or other third party agreements, with a portion of the proceeds returning to Broadjam. At the time, companies like mp3.com were giving away and selling music to consumers, as well providing an ad hoc distribution mechanism to music placement opportunities for radio, TV, and film.

Just as quickly as the chaos emerged, however, legal governance and corporate control mechanisms were reasserted. The major labels, in conjunction with big name bands including Metallica and Dr. Dre, sued Napster. The RIAA filed lawsuits against large ISPs and even major universities to obtain the identity of illegal downloaders. The RIAA subsequently sued thousands of individual consumers, despite widespread criticism and poor publicity. Although mp3.com was sold in 2001 to Universal Vivendi for $372 million, the mp3.com model of offering free music couldn't be monetized in the face of the recording industry's legal clout. The assets of mp3.com were sold to CNET in 2003 for an undisclosed but significantly reduced sum. Napster lost its lawsuit against the RIAA in 2002 and was sold to German media giant Bertelsmann, but the sale was ultimately invalidated in court. Napster was shut down.[1]

The implications for Broadjam were grim. By 2003, the distribution channel Broadjam anticipated utilizing to monetize its catalog had effectively been litigated out of existence. But eight years later, Broadjam serves more than 90,000 independent musician customers, places hundreds of their works into a variety of commercial opportunities, and holds a catalog of more than 300,000 unique songs. While the company isn't publicly listed and may be unknown to many music industry insiders, it accomplished something that almost no other online music sites from the Napster-era managed: it survived.

How?

INSIGHT 2: APPRECIATE IMPERFECT OPPORTUNITIES

The second insight to achieve the unexpected is to acknowledge and even value imperfect opportunities. There are two important implications to this insight. First, successful entrepreneurship often requires accepting the limitations of objective analysis. Entrepreneurs must make do even when opportunities turn out to be flawed, misunderstood, or only partly accessible.

Second, even when good data and objective assessment are possible, they may not be enough. Opportunities are sometimes imagined to exist in a perfect state to be identified and fully understood by the alert entrepreneur. While entrepreneurs surely must be alert to opportunities, it seems unlikely that the most interesting opportunities can be fully understood without actually starting the commercialization journey. We have repeatedly heard entrepreneurs describe opportunities evolving over time. The most extraordinary entrepreneurs shape opportunities to their own purposes. It appears that some entrepreneurs achieve the unexpected precisely when opportunities are unformed or imperfect. Perhaps if all opportunities were perfect, outcomes would be inevitable and predictable. Elkins may have believed the internet distribution opportunity linking independent musicians to customers was perfect, but by 2003 he had no illusions about flaws in the business model. But eventually it would emerge that the flaws that doomed many online music distribution companies would eventually be the leverage points for Broadjam's success.

OF SPHERES AND SHARDS

The entrepreneur who will compete in this expanding, hypercompetitive world of opportunities should understand that *opportunities are shards, not spheres.* Take a moment to consider a joke, ostensibly about differences between various fields of science. Granted, it's not an especially good joke, but we believe it effectively invokes the inherent imperfection in opportunities.

> A chicken farmer, who currently plucks chicken feathers laboriously by hand, imagines a machine that would do this task quickly and efficiently. Without the knowledge to design it himself, he hires a mechanical engineer to design the machine. After months of research, the engineer gives up, and recommends an ornithologist. The ornithologist, after further months of effort, also gives up and recommends a physicist. The farmer is skeptical, but contacts the physicist to explain the idea. To the farmer's surprise, the physicist arrives at the farm the next morning with the solution drawn on blueprints.
>
> "But the other two said it was impossible!" cries the farmer.
>
> "It's no problem at all." replies the theoretical physicist. "First, we'll assume that a chicken is a perfect sphere with uniform density and zero mass ..."

Imperfect and irregular opportunities

Scholars as well as the popular press often discuss opportunities with an explicit assumption that they can be well-identified, isolated, and targeted. The idea that opportunities are, in effect, perfect spheres, is attractive but unlikely. Perhaps, with more sophisticated tools than are currently available to students of management and entrepreneurship, opportunities could be broken down into perfectly specified components that can then be analyzed. This is somewhat akin to arguing that because chickens are made of molecules, there may be a way to defeather them at the molecular level. It certainly makes for interesting conversation, especially if you are half-farmer, half-theoretical physicist. And it is an equally interesting, albeit long-term research question for those of us who study entrepreneurs. But the entrepreneur herself is probably less intrigued, preferring instead to just get started lest someone else get there first. Opportunities simply don't look like spheres in the real world. They are shards: imperfect and irregular.

What does this mean? Why should entrepreneurs be interested in imperfection?

Return briefly to our hapless chicken farmer – entrepreneur. He sees a relatively simple opportunity: if he can reduce his variable operational costs, he'll make a higher profit. This hardly qualifies as business model innovation, but let's start simple. The farmer has assessed the situation, identified the goal, and tries to access the resources required to reach that goal. In the process, however, he has learned that while there are many resource paths to the goal, not all of them *make sense*; even the ones that are accurate within a strictly logical framework. This may seem trivial, but we see this happen all the time in entrepreneurial contexts.

Let's bring this into a more traditional organizational context, entrepreneurial or not. An opportunity comes to the attention of the manager(s) of a business. The opportunity is evaluated by reasonably sophisticated businesspeople who analyze markets, competition, and finance. They consider the resources and match them to expert assessments of what will be required to address the opportunity in terms of technology development and logistics. They've read books on strategy, operational excellence, competencies, and business models. They put all of this together, and launch the project. The project may fail for unrelated reasons, or it may be successful. But whether it succeeds or not, it may be to the detriment of the company, right away or in the long run.

Why? First, really interesting opportunities aren't spheres – and they can't be perfectly identified or even broken down into perfectly identifiable parts. This isn't a case of the blind men incorrectly identifying the various bits of an elephant; it's a case of sharp-eyed people with sophisticated tools attempting to isolate and categorize something that may be understandable *only after the fact*. An elephant is an elephant – step far enough back and you'll recognize it. But there simply is no guarantee that a given opportunity can be accurately and fully characterized, regardless of how many perspectives or lenses are applied. New opportunities are shards, almost like fractals – irregular parts of a larger whole that is unknown and possibly unknowable. It's not unreasonable to think that the firm could develop a relatively clear understanding of many characteristics of a specific opportunity, but placing that opportunity in the broader market and industry context adds yet another level of complexity. It isn't necessarily impossible, but the more complex the individual factors associated with the opportunity, the less likely that more sophisticated modeling will clarify the situation.

Malcolm Gladwell provided an example of this problem to demonstrate the power of intuitive, decentralized analysis. He described a US military simulation in 2000 intended to leverage massive advances in computational capacity for assessing combat situations and managing complex activities.[2] In practice, the information collection and assessment handicapped the analytical process. The decision heuristics required so much information, and processed that information so slowly, even with supercomputer technology, that central command couldn't keep up with the rapidly changing situation on the ground. This is not the same as "analysis paralysis" in which information glut prevents decisions from being made. That paralysis results from indecision on the part of individuals faced with more information than they can realistically understand and assess. The situation described by Gladwell is characterized by analytical processes that cannot operate at the necessary granularity or generate results in a timeframe relevant to real-world rates of change. Decisions will be made, but with potentially flawed or out-of-date data. In this way, opportunity shards resist perfect analysis.

We must remind ourselves, once in a while, that organizational design is not a perfect science. Aligning an organization to tackle a novel opportunity is not just a question of breaking a business down into its component parts and realigning those parts to "fit" the opportunity. Organizations are not engineering problems that can be reduced to a set of linked equations.

The benefits and drawbacks of opportunity spheres and shards can be understood in the context of business model dimensions. Perfect opportunities result in clearly defined business model dimensions, but often lead to inflexible solutions that may be easily replicated in the market. Imperfect opportunities require organizations to test, create, and adjust. This could help entrepreneurial firms find entirely new sources of value creation, but could just as well lead to dead ends or solutions that can't be scaled. Table 2.1 compares the pluses and minuses of opportunity spheres and shards across the business model dimensions.

Imagine, for example, that the firm is a video game developer specializing in 3-D imaging and behavior of vehicles and robotic systems. These competencies would lend themselves towards the development of simulators for military or other government applications. At a time of rapid defense spending increases and limited manpower, the US military has actively sought novel ways to utilize technology to reduce manpower training times and long-term casualties. Consider, for example, the General Atomics MQ-1 Predator drone. Could our video game developer apply its skills to service this market?

The answer is yes, but as with our chicken farmer, it's possible that the way we are addressing the opportunity is, in fact, the problem. Unlike the chicken farmer story, however, this one is real. The firm in this case is Savage Entertainment, a successful independent game production company based in California. Started by two former Activision employees in 1998, Savage has produced or

Table 2.1 *Perfect imperfection*

Business model dimensions	Spheres (Fit)		Shards (Imperfection)	
	Good	Bad	Good	Bad
Resource structure	Perfectly adapted to activities	Inflexible, valueless outside of specified use	Unique value for your organization and opportunity	Difficult to scale
Transactive structure	Well-specified, scalable	Routinized and uninteresting for participants	Challenge encourages experimentation and adaptation	Hard to replicate, slow and tacit learning processes
Value structure	Clearly identified, can be communicated to investors and stakeholders	Obvious to competitors, subject to vertical integration, accessible via finance, approximations may be far off mark	Difficult to copy or access, special value to certain, possibly large, segments	Trial and error may be required, false signals from customers and industry because of unfamiliarity or uncertainty in needs

supported production of some of the top video games of the past decade. Savage developed some of the highest profile "properties" in the business, including *Transformers*, *Star Wars*, and *James Bond*. In fact, *Star Wars: Battlefront 2* is the second highest selling game all time on the Sony PSP. The firm's work on military projects was successful enough to get Savage nominated for a DARPATech award in 2007. But in many ways the decisions to target those opportunities were stepping stones in a process already under way at the company that led to serious problems. Simply put, the opportunities available to Savage were imperfect.

Like most entrepreneurial firms, Savage suffered from resource constraints, organizational structure limitations, and limited information and rationality. Savage was run by two extraordinarily intelligent individuals, Tim Morten and Chacko Sonny, both with extensive management and industry-specific experience. Both were avid and sophisticated consumers as well, attuned to the changing interests of customers and broader market shifts. Every executive decision was rational and, by most management theories, well-planned and implemented. In the end, however, it turned out that at some point in the growth of the organization, good decisions were not cumulative, and the route to success did not lead through a series of optimal decisions. Savage's decision to pursue military contracts was in line with traditional tenets of strategic management. The projects leveraged and extended core competences, contributed to firm profitability, provided human resource stability within the notoriously cyclical game development process, and engendered leading-edge legitimacy for the firm. The "fit" was excellent. This was, in fact, the problem.

The question that needed to be asked wasn't whether, from a strategic perspective, the organization was well-fitted to the opportunity. The real question deals with the long-term implications of contract work given the founders' long-term plans and interests. This may not have been a failure of business strategy, but whether diversification of contract work represented a viable component of a sustainable business model.

Contextual relativity

Researchers have shown that opportunities look different to different people. Experience and expertise form the primary contextual elements in addressing novel opportunities,[3] but not the only ones. Entrepreneurs and managers must assess industrial and organizational context to ensure they ask the right business model questions. *Contextual relativity* means that business models address opportunities within these contexts, not in a vacuum. The same business model applied to the same opportunity, but within different organizational context, won't generate the same results.

Let's look again at the situation with Savage, and apply the concept of contextual relativity at all three levels to see what we can learn.

At the level of the opportunity itself, the managers at Savage were torn between their original vision and the operational realities of the firm they were running. Sonny and Morten had left Activision to build entirely new "products": novel and creative games. But their first project was actually contract work for Activision negotiated as part of their departure from the firm. In the first years, Savage focused on one project at a time, but a financial crunch caused by a non-paying client led the Savage entrepreneurs to reconsider the one-team/one-client model. Diversifying the company's activities reduced reliance on any one project or client, but took the organization further from the original opportunity: developing novel gaming content. Military-related projects presented a different but related opportunity, as did contracts to port content from one platform to another.

At the organizational level, Savage appears to have continuously adapted structures and activities to address the perceived opportunities. The original one-project structure gave way to functional structures reflected in both project design and even the layout of an expanded facility. This enabled competency groups, such as artistic design and engineering, to support multiple projects while simultaneously maintaining stable and capability-enhancing subgroups.

One critical note associated with organizational context is that the founders, Sonny and Morten, remained actively involved in game development – including aspects of look and feel – and even, at times, coding. This reinforced a "job shop" organizational narrative that matched the opportunities addressed at the corporate level.

Next, we have to understand what happened in the firm's industrial context. When Sonny and Morten left Activision, the video game development sector was in a repeated cycle of changes in platform dominance, often based on the success or failure of a single game. In 1993, Nintendo had 90 percent market share, but Sega leapfrogged Nintendo when it introduced the NES system and the Sonic the Hedgehog game, capturing 65 percent of the gaming market at the time. That success was relatively short-lived, dropping to 35 percent by 1996. By 2001 Sega had abandoned its hardware platform altogether. Since then the hardware wars have continued unabated, pitting Nintendo against "newcomers" Microsoft and Sony. Video games remain one of the fastest growing and most profitable entertainment segments. In 2005, video games surpassed total movie revenues in the United States. Revenues have nearly doubled to roughly $20 billion in 2009, far outstripping movies and music. Video games represent the largest global entertainment mode; DFC Intelligence projects that the global video game market will grow from $65 billion in 2011 to more than $85 billion by 2016.[4]

Viewed from the outside, the video game industry appears to be a consistently profitable, growing sector. But from the inside, the past ten years have been turbulent and challenging. First, development costs for a major game release have increased dramatically. In 2000, the cost of designing and producing a top quality video game ran between $2 million and $5 million. Peter Molyneux's god-game, *Black & White*, a genre-defining accomplishment that took three years to complete, was considered exorbitant at $6 million in 2001.[5] But by 2009, the cost of producing a "hit" video game had skyrocketed to $20–30 million.[6] Multiple factors were at play in this escalation – the ever-accelerating capabilities of personal computers and game consoles, the requirements for remote, multiple-player play, and the growing dominance of so-called legacy products: sequels and entire series based on a single "world" or conceptualization. The most recent major installment of *Grand Theft Auto*™, *GTA IV*, had an estimated production cost of $100 million, exceeding the budget for all but the biggest Hollywood blockbuster films.

Why is this important?

The video game industry has been shifting from a model in which novel "properties" came primarily from independent developers to one dominated by major studios that can afford the massive investment in distribution associated with blockbuster hits. Like the venture capital and American movie industries, video gaming developed a returns profile in which a few big hits make up the vast majority of revenues and profits.[7] The hurdle for new product development has escalated far beyond the capabilities of a few creative die-hards working in a garage or warehouse – the rapid advances in CAD, rendering engines, and raw processing power work against entrepreneurs who want to design new games, because the sophistication of animation and gameplay has scaled so rapidly. We want to believe that the real value in the development process is the creativity of the original designers, but the scaling of the development, marketing, and sales processes weights the odds in favor of established properties, and big studios have invested to reflect this new reality.[8]

The final component of industrial context has emerged in the past five years. Even as the industry grew, the major studios struggled to scale with the increasing complexity of game production. This is a critical issue for an industry that has two major marketing and sales periods: summer and Christmas. A missed deadline can be the difference between an unprofitable property and a blockbuster. Initially the studios relied heavily on outsourced talent, contracting with firms like Savage for legacy products, especially as a stopgap in time-constrained development situations. But as these costs increased and long-tail effects grew more pronounced, the studios began buying up capacity to gain better control of production processes.

By 2009, Savage was a rarity, one of the oldest independent game developers employing more than 50 full-time staff. The big players, Activision and EA, had been on buying sprees:

> In the past five years alone, Electronic Arts has acquired developers JAMDAT Mobile, Mythic Entertainment, Phenomic Game Development, Digital Illusions CE, Headgate Studios, and VG Holding Corp. And it controls 15 percent of Ubisoft. All told, the acquisitions cost the company billions of dollars. For its own part, Activision has been just as active. The company acquired Vicarious Visions, Toys for Bob, and Beenox in 2005 and followed that up in 2007 by acquiring a controlling stake in Bizarre Creations, which was trailed by its 2008 merger with Vivendi to become Activision Blizzard.[9]

The financial crisis in 2008 and 2009 dramatically affected outsourced development, even as video game sales continued their relentless climb. Studios and distributors cut back on investments in new properties and delayed launches to reflect the broader economic recession. Production contract opportunities fell dramatically, because the studios had more capacity than required for the reduced level of new development.

Finally, remember that the reason the Savage founders left Activision in the first place was because they wanted to design and develop original video games based on their own efforts and ideas. This is the entrepreneurial context or entrepreneurial narrative that drove the creation of the firm. It is important to understand that entrepreneurial narrative can change over time. We can see this very clearly in some of our other study companies like Praj Industries and Metalysis, and in less obvious shades at companies like Confederate. Changes in ownership and executive management may be either leading indicators or reflections of narrative shifts. But at Savage, where the founders remained both executive management and owners, the change in organizational activity did not reflect a change in entrepreneurial narrative. The managers, and their employees, believed that the contracting work was ultimately secondary to a long-delayed vision for developing entirely new games. In reality, the company had long-since become, as Sonny described it, "the go-to garage for high-quality, fast, and efficient outsourced game development." When the recession effectively eliminated that kind of contract work, Sonny and Morten briefly looked at mechanisms to jump-start development of new game content. They quickly realized that they simply couldn't get access to the funds they would need, given their lack of a track record in novel content development. In spite of their exceptional effort and initial success, they could not realize their original intellectual property development goal, despite ten years of functioning as a contract developer. The business model in place at Savage had effectively evolved away from the founders' original narrative, and was incongruent with the evolving, imperfect opportunity.

By late 2010, Savage looked like a completely different business. Staff had been reduced to a half dozen programmers and designers working on ancillary projects related to prior work. Tim Morten stayed on to oversee completion of active projects and seek an acquirer for the firm's assets and book of business. Chacko Sonny followed the trend in the industry and returned to Activision, where he is building a new studio focused on developing a new social platform and on-line service for the *Call of Duty®* franchise.

Hindsight is 20/20, of course. We can't say what would have happened without the global financial crisis, but our observations of Savage a full year before the contraction of the video game development outsourcing space suggest that the mismatch in entrepreneurial narrative and the contours of the evolving opportunity was already a significant factor for both founder-managers.

Seeing shards

Contextual relativity means that assessing opportunities necessitates observational lenses, but it also means that there are multiple possible

Table 2.2 *When perceiving opportunities as spheres or shards is most valuable*

Spheres are valuable when ...	Shards are attractive when ...
– Costs of development are rising faster than market growth – Specialization is necessary to compete but commoditized in at least one key factor market (such as labor) – Missing the market by a little means missing it completely – Technological or economic shifts are common and/or likely to benefit large, diversified firms	– No-one is quite sure what the market wants (shaping potential) – Technological or economic shifts benefit knowledgeable participants (non-competence destroying) rather than well-resourced participants – Near misses generate interest – The organization learns faster than it can implement

enactments of the same kinds of opportunities. It's all about seeing shards and appreciating idiosyncrasies rather than imagining perfection. Table 2.2 shows the general conditions under which appreciating opportunities as spheres or shards are likely to be most valuable. Let's examine opportunity shards in a specific area and look at two dramatically different approaches to seemingly similar problems.

The world of human limb prosthetics can be both depressing and uplifting. Prosthetics are imperfect solutions to horrific traumas, whether suffered via disease, violence, or accidents of genetics and birth. At the same time, prosthetic technologies can be triumphs of empowerment and spirit. From a business perspective, prosthetics are a multi-billion dollar industry where the largest customer markets have the least ability to pay for innovative and high-functionality products. In general, technological advances significantly increase cost of development, regulation, manufacturing, and distribution.

One approach to these challenges is exemplified by the extraordinary work being done jointly by the US government, specifically DARPA and the Department of Veteran Affairs, and DEKA Research and Development, Dean Kamen's innovation business. The DEKA Arm, dubbed "Luke" after the *Star Wars* character, is considered one of the most advanced human-robotic interface devices ever created. According to Dr. Stewart Coulter, Project Manager at DEKA Integrated Solutions Group, at least $30 million has been invested in the technology since 2007:

> We took on this incredibly difficult challenge from DARPA. Essentially, we were asked to start with a clean sheet of paper and develop a prosthetic arm system fitting within the size and weight of the original equipment that would have the dexterity to enable the user to pick up a grape without crushing it, and the arm and grip strength to pick up a gallon of milk without dropping it. It would need the sophistication and flexibility to run a power tool, reach a high shelf or whatever else they wanted to do. And, of course, it would have to do so with a non-invasive, intuitive control scheme so recipients could start using the arm quickly.

The device uses modular electronics, flexible circuitry and microchips, sophisticated tactile feedback systems and novel mechanical attachment mechanisms to limit discomfort. At this time, there's no indication of what production costs might look like, but DEKA's goal was to produce a robotic arm that would provide full functionality.[10]

An equally radical approach to a slightly different problem can be seen in ReMotion's JaipurKnee project. After starting as a joint project of Stanford's Mechanical Engineering department and BMVSS, a non-profit organization that is the world's largest provider of leg prostheses, ReMotion continued on to develop a viable knee prosthesis that costs less than $40 to manufacture. Inspired by the JaipurFoot, a $40 artificial foot that has been provided to nearly one

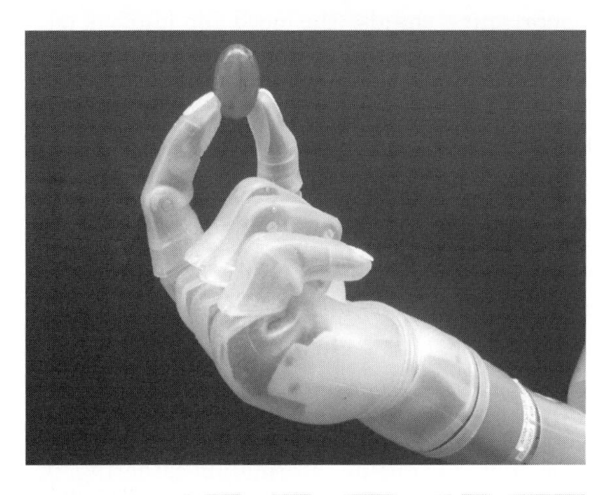

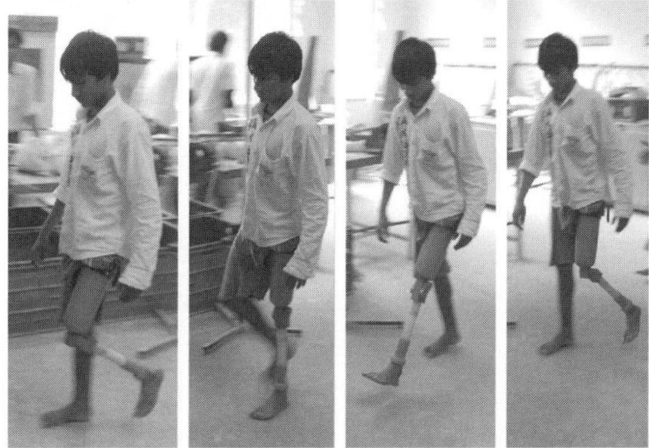

FIGURE 2.1 Entrepreneurial realizations of opportunity shards
Images courtesy of DEKA Integrated Solutions Group and ReMotion Designs

million people worldwide,[11] the JaipurKnee was named one of *Time* Magazine's 50 best inventions of 2009 and is worn by thousands of people in India and other countries.[12] Figure 2.1 shows the DEKA hand and the ReMotion JaipurKnee in action.

It would be easy to focus on the difference between an arm-hand prosthesis and a knee-ankle prosthesis to justify different approaches to opportunity realization, but the scale of investment and target cost difference between the projects represents a much greater philosophical gap in how to address an opportunity. These approaches reflect opportunity shards: the need for idiosyncratic prosthetic devices. The total market may be identified with some certainty, and segmented into specific shards such as impoverished rural villagers in India or American soldiers. At the same time, the broader trends in health provisioning worldwide suggest that such clear divisions and classifications are beginning to fade.

Globalization processes are eroding barriers to markets and producers; intercommunity involvement has been massively democratized by YouTube while global-scale financing has been democratized by online payment systems epitomized by PayPal. Even as market segmentation tools become more sophisticated, with access to more and more private data via online information gathering, the underlying reality of those markets is shifting faster. For instance, how different is the market for mobile ringtones in Tokyo and New York? How long will those differences last?

Entrepreneurs will always face the challenge of identifying extraordinary opportunities. But in the coming years, as opportunity identification and access present a more level and global playing field, entrepreneurs will need to use organizational design tools to effectively pick out shards and realize their value. In that process, entrepreneurs must appreciate the difference between familiarity and knowledge. We are only now beginning to appreciate how familiarity affects not just our interpretation of specific opportunities, but actually filters and influences thought and analytical processes. Early research on this fascinating effect, called cognitive fluency, demonstrates that a sense of familiarity generates a corollary perception of proximity. In other words, prior experience and knowledge influences the type of opportunity that entrepreneurs perceive, the mechanisms and processes they apply to exploit that opportunity, and even the entrepreneur's assessment of the length and difficulty of the exploitation process.

The insight of appreciating imperfect opportunities goes beyond opportunity identification and assessment. Why? Because our theories about how organizations succeed are based on frameworks of strategic fit, that make unreasonable assumptions about managing constraints. Just as opportunities are not perfect spheres of uniform density, organizations aren't perfectly organized collections of well-defined resources and activities. This may seem obvious, but to appreciate why we have to adjust our view of entrepreneurial management, we have to start with the long-standing and well-respected framework of strategic complementarity.

LIVING WITH CONSTRAINTS

The theory of strategic complementarity was developed by Professors Paul Milgrom and John Roberts, two insightful and influential economists. Milgrom and Roberts, building on new theories of resource complementarity,[13] applied economics theory and case studies to show that organizations would outperform by managing internal activities to generate enhancing rather than conflicting characteristics.[14] In other words, organizational effectiveness requires adapting resources, people, and processes to create mutually reinforcing systems that optimize firm efficiency as fitness within a given environment. A business is therefore only as strong as its weakest link. Weak links need to be culled, patched, or somehow inverted to ensure the system is fully consistent; building on the strongest links strengthens the organization.

This is an intuitive and powerful idea. Like many important business insights, it blends common sense with a strong basis in economic theory. For the past 20 years, the success of numerous businesses has been retroactively explained with this analysis. Let's take a quick look at a couple of examples.

One often cited example is Southwest Airlines, a poster child of sorts for competitive advantage based on the effective implementation of a low-cost strategy. Figure 2.2 shows a well-publicized graphic that demonstrates the consistency across Southwest's organizational processes. It appears that all of the operational and strategic elements of the organization are mutually enhancing. This explanation of operational excellence is compelling. Intuitively, any organization

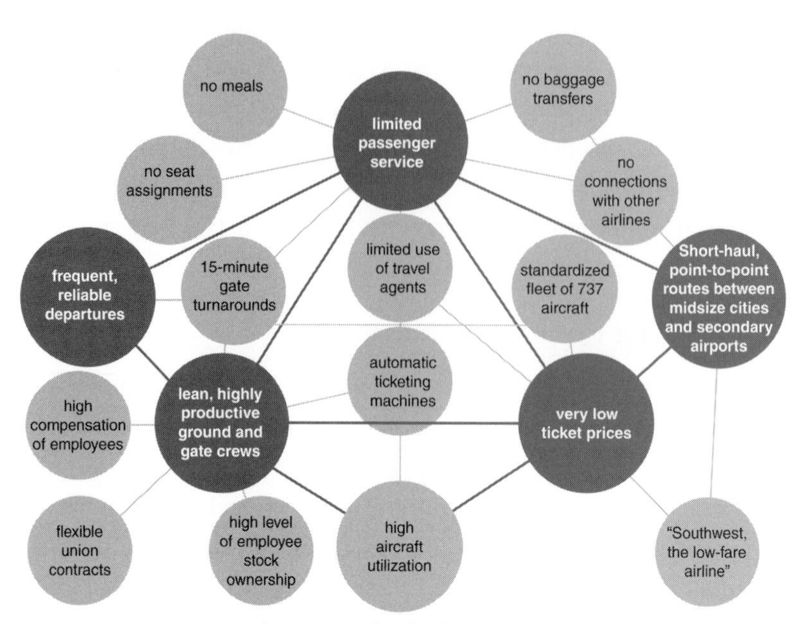

FIGURE 2.2 Southwest's perfectly aligned activity set
Adapted from Porter, M. E. 1996. "What is strategy?" *Harvard Business Review*, 74(6): 61–78.

would seem to be most effective when all of its people, activities, and interactions are mutually reinforcing. From a strategic perspective, the firm demonstrates internal and external "fitness." Internal operations function at maximum efficiency; transactions with partners are presumed to be made efficient, and this optimization makes the firm more competitive in its environment than firms that demonstrate inconsistent systems. Southwest very effectively exploits the opportunity to provide low-price air services as the world's first and arguably most successful low-cost air carrier. Certainly the financial results are unassailable: while the airline industry lost nearly $60 billion from 2000 to 2009,[15] Southwest continued its run of profitability to 36 years in a row.

Given that Southwest is now a highly resourced organization operating on a long-scale timeframe with relatively clear operational goals and priorities, it isn't unreasonable to see this as a critical component of the success of the organization. On the other hand, it isn't difficult to poke holes in aspects of this explanation. For example, "high compensation of employees" appears to be only a partial match with "lean, highly productive ground and gate crews." Clearly, high pay may be linked to high productivity, but the chain of coherence from "high compensation" to "very low ticket prices" is not perfect, especially as the strategic positioning of the organization is a low-cost competitor.

In fact, Southwest has focused on creating a high-quality workplace that values employees. From a cost perspective, Southwest has worked to ensure that high-cost items like health insurance are both reasonable and efficient, but there is no shortage of anecdotes about decisions that challenge the certainty of the perfect low-cost activity set.

> While other airlines announced they were cutting their workforces at least 20 percent in the wake of the post-9/11 drop in business, Southwest kept all its employees on its payroll, and even went ahead with a $179.8 million profit-sharing payment to employees three days after the attacks.[16]

The argument that all of this is part of a long-run plan to keep costs low sounds good, but how far can we stretch that logic? With that in mind, consider the following statement:

> I think we saw the financial crisis coming, maybe better than other companies, and in fact our revenues showed some growth while everything else seemed to be contracting. Not nearly as much growth as we had projected, of course, and we'd been building infrastructure and hiring based on those growth projections. But we made a commitment to our employees – I actually called a meeting of the whole company to tell them – we were going to defend the human capital of the organization and do our best to make sure everyone kept their jobs, their compensation, and their benefits as we braced ourselves for the storm.

Is this Southwest again? No, it's Matt Blumberg at Return Path, talking about a similar commitment to employees. Here's the problem – Return Path is not a low-cost company – in fact the company charges significant premiums for its differentiated, leading-edge products and services. Blumberg does want to run an efficient company, but he is certain that Return Path's success will come from being the market leader in technology and capabilities, not cost management for the sake of keeping prices as low as possible for a minimum service level.

In other words, two companies with radically different strategies are successfully using human resource models with strong similarities. But if that's the case, then the argument that both have perfectly enhancing activity sets simply can't be true. There are many ways to explain this result. Let's look at a number of factors to set the stage for what really happens inside firms and how entrepreneurs leverage internal inconsistencies.

Subscale effects

First, internal consistency is more important at larger firms, while entrepreneurial firms can survive and even thrive with internally inconsistent elements. Our research suggests that while internal consistency clearly has benefits, when firms are "subscale," those benefits may be limited. In special cases, in fact, there may be entrepreneurial advantages to maintaining apparently conflicting elements within an organization. For example, let's consider the case of Voxel.net.

Voxel dot Net (Voxel.net) is an internet infrastructure company. In its own words, it provides scalable hosting and delivery of online content. It hosts websites and related content for high-bandwidth clients. It was one of the first cloud-computing companies, and its network was recently identified as the fastest in the US and one of the fastest in the world.[17] Amazingly, this company of less than 100 employees grew organically without venture funding and yet competed against big, branded organizations like Akamai and Rackspace, defying traditional growth and strategy theories.

The company was started by Raj Dutt in his dorm room at Rensselaer Polytechnic Institute. Back in 1999, Dutt was just reselling hosted server space, brokering capacity on other companies' servers. At the time, network administration knowledge was quite specialized and profitable, and Dutt found himself with more business than he could handle as a student. Ultimately he left RPI before graduating to build Voxel.

From these humble beginnings to 10GB interconnectivity everywhere via nearly 20 global nodes, Voxel had yet to bring in venture capital, so-called "professional management," or even outsourced customer support. In fact, Dutt traces the success of the company to the fact that customers needing help reached knowledgeable engineers immediately without going through an administrative filter. At the same time, in order to compete with companies like Amazon, Akamai, and Rackspace, who generate billions of dollars of cloud-based revenues in an industry with incessant price pressures, Voxel has remained price competitive. *Voxel was using differentiated services as part of its low-cost strategy.*

As we noted, however, this might not be scalable. The small size of the company means that most of the market-facing employees are in close proximity to each other, both physically and in terms of fundamental knowledge. So solving customer problems can rely on access to the engineers who set up the customer in the first place. Significant growth could make this less efficient. Over time it's likely that the high-touch model would accrue higher relative costs. One explanation for successful use of non-enhancing elements at entrepreneurial firms is simply that they are subscale.

While reasonable, this explanation can't be the whole story. Entrepreneurial firms are generally subject to *more* pressures and challenges than large firms and more likely to fail, a phenomenon scholars refer to as "the liability of newness."[18] If the presence of conflicting elements is inherently problematic, entrepreneurial firms with internal consistency issues should, on the whole, be even more failure prone.

Dynamic effects

Another aspect of living with constraints is the plastic nature of activities and relationships within an organization. Organizations are dynamic systems operating in dynamic environments. Changes in personnel, technology, routines, and information affect how the organization functions, sometimes predictably, sometimes unpredictably. Managers may leverage these processes to improve how the organization operates and competes. But not every change can be predicted, much less controlled. Sometimes management's response to change appears to be optimal but actually leads to longer-term challenges, as in the case of Savage.

There are two levels of dynamic change that managers address in this context. The first type puts the extant set of relationships out of balance. Imagine, for example, the impact of nationalized health care on human resource management at Southwest. As an alternate example, imagine that the company develops a novel solution to baggage handling that changes how many people are required to prepare a plane for departure. In either case, the importance and relationship between activities, resources, and perhaps even internal priorities will be altered, and the enhancement or conflict between elements could change, requiring adjustment. At Savage, a military client might prefer a different balance of artistic creativity versus user control. Management might respond by adjusting the number or qualifications of design and engineering staff associated with the project.

The second level of change is systemic. These are environmental or internal changes that affect the overall value of the firm's configuration. Southwest might decide to become carbon neutral by a given target date or set a goal of 95 percent customer satisfaction. Savage might have decided to only develop products for the government, or edutainment software, or the owners could have decided to re-form the company as a cooperative. In these cases, the old configurations would have significantly different overall value because the measurement criteria have changed.

Regardless, creating and assessing *a static configuration* is only useful *in static environments*. Some industries are relatively stable, but many are not, including most new, entrepreneurially driven industries. In other words, "fit" is necessarily a static concept, while most organizations operate in a dynamic world.

Complexity

Information processing places constraints on managers' abilities to assess, control, and change internal systems. We might expect entrepreneurial firms, which tend to be smaller, to be less subject to information processing complexity. On the other hand, these firms operate in less well-established industries, and employees often execute a wide range of activities, including experimentation with new activities. More than half of the employees we interviewed at our case study companies told us that experimentation was a regular, if not expected process for getting *day-to-day* work done. In fact, they told us that experimentation was "as" or "more common" than explicit managerial control for the firm's *critical* activities. Creating a clear picture of firm activities when half of those activities are experimentally determined obviously presents potential challenges; assessing such complex configurations under high levels of uncertainty presents further challenges. *Optimizing* perfect configurations presents a high bar, indeed.

The good news, as we'll discuss in Chapters 3 and 4, is that organizational configurations can be assessed without enlisting a supercomputer. We simply don't know whether the level of organizational complexity, or management's ability to assess complexity, affects the relationship between internal consistency and competitive advantage. Our observations of entrepreneurial firms suggest that optimization simply isn't the dominant heuristic in how early stage firms evolve and address opportunities.

Multilevel goals

One of the surprises we encountered in our case studies was that the so-called double (or even triple) bottom line has been replaced, whether explicitly or implicitly, with series of multilevel goals. At the same time that inherent exogenous complexity would suggest that entrepreneurs focus on simplicity, the entrepreneurs we talked to seemed to be making life more complicated.

Let's look at two examples before we define types of multilevel goals. Understanding the types of multilevel goals will be especially important when we consider how entrepreneurs balance constraints and use complexity to benefit organizational stability.

Talk to Matt Chambers, the founder and CEO of Confederate Motorcycles for more than two minutes and you're likely to find yourself enmeshed in a philosophical critique of modern capitalism and business practices in a broader context of human purpose. When he says, "I don't need this business to make money to be successful," he means it. This may sound strange, ironic, or downright oppositional coming from the CEO of a 12-person manufacturing company that's listed on a public stock exchange. But Chambers is serious – intensity and passion underscore every word spoken in his gentlemanly Southern drawl. And Confederate Motorcycles is not your average manufacturing company: how many companies sell $85,000 motorcycles to people like David Beckham, Brad Pitt, and Tom Cruise?

Chambers isn't exactly proud of the fact that Confederate has generated relatively low profits in its 15-year existence. But considering he's funded the company – twice – with his own money, his commitment can't be questioned. While he plans for the company to generate profits and succeed commercially, the organization and its products accomplish other goals. Chambers describes how purposeful design taps into emotional response:

> If you are from the [American] South you appreciate the sorrow of living in conquered states. I'm not talking about slavery and

> freedom or plantations and industrialization, or right and wrong. But there's a sympathy, an empathy of being from the South that I want to express, and I'm expressing it in the medium of industrial design. Lots of current cycles have all these extra parts to look smooth or complete ... but the Wraith has empty spaces purposefully designed into it, like a skeleton, the bones of the South, and evokes that Southern sorrow, even for people who aren't from the South. My mission is to immortalize that in the machinery that will never fatigue.

There's no doubt that Confederate has had an effect on motorcycle design, and perhaps industrial design more broadly. Larger manufacturers, including Indian Motorcycle and Aprilia, have introduced concept motorcycles that embrace some of the Wraith's design elements. *ID* Magazine named Confederate's B120 Wraith the "World's Sexiest Motorcycle."[19] The Wraith also holds the world land speed record for its engine class and won first prize for production motorcycle at the 2007 World Championship of Custom Bike Building in Sturgis, North Dakota. By contrast, Confederate's P120 Fighter was featured on the cover of Neiman-Marcus' Christmas catalog. Chambers has found a way to manage multilevel goals.

Recurve, Inc. provides a much simpler example. Formerly Sustainable Spaces, this San Francisco-based company started out as a specialty consulting and construction company. Founder Matt Golden saw that huge gains in energy efficiency could be made in residential homes with targeted, customized remodeling. But as he built the company, he saw that a specialty construction business would be nearly impossible to scale, partly because of dramatic regional differences in construction practices and requirements. His personal goal was to have as large an impact on energy use as possible. Now Recurve is rolling out software to enable other construction companies to perform the award-winning home audits and remodeling work that made the company famous in the Bay Area.

To make this transition possible, the company brought in an experienced CEO and took venture funding. Have those changes now limited the organization to the classic venture capital goal of profitability?

Incredibly, this isn't the case. On the one hand, Golden's goal of maximizing impact fits well with venture capitalists' expectations – scale and value creation are an excellent match. However, Recurve has retained a variety of organizational practices that would make someone like "Chainsaw" Al Dunlap cringe. The company orders all of its office supplies through a broker that guarantees green sourcing and even donates a small percentage of revenues to ecologically sustainable causes. The company's human resource policies are designed to focus on "sustainable development" of employees, including generous benefits and a dedicated focus on a positive culture. In fact, the company has a chief culture officer rather than a traditional VP of human resources.

One could argue that for Recurve these policies are a matter of survival operating in San Francisco and Silicon Valley. After all, Recurve is now competing for talent with software companies rather than construction firms. If that's the case, of course, and such policies are inherently antithetical to venture capital and value maximization processes, we would expect to see an exodus of software and related businesses from the San Francisco area to lower-cost cities. While we do see outsourcing of a variety of business functions, and there are cases of corporate emigration or at least outgrowths to other cities, there is little doubt that, for now, the Bay Area and Silicon Valley remain a powerful magnet for highly innovative firms, even firms coming from venture hubs like Boston. Recurve isn't the anomaly. If the Bay Area is the anomaly, then we may have to expect that this anomaly, like so many other trends and philosophies, can be expected to spread from California to the far corners of the globe.

Innovative entrepreneurs and organizations are establishing and meeting *multilevel goals* as integral parts of operations and definitions of success.

Our studies suggest that multilevel goals demonstrate common characteristics, and understanding these characteristics is one of the keys to developing highly effective organizations. Obviously, goals can be implicit or explicit, simple or complex, proximal or stretch. These dimensions influence operational management and leveraging resources, but there are two other goal characteristics that seem to fundamentally determine how entrepreneurs build business models and live with constraints.

The first key multi-goal characteristic is aspirational locus. Multilevel goals may be associated with generating results at individual, group, organizational, industrial, geographical, or even social levels. ValueLabs, the Hyderabad-based software company, has explicit policies about maintaining a positive work environment, including access to natural light. Recurve wants to drive a national debate about residential energy efficiency. Confederate wants to influence industrial design to reflect a humanistic philosophy.

The second key characteristic is organizational commitment. When entrepreneurs develop multilevel goals, the level of organizational commitment to those goals is a critical indicator of how the company must be structured to accommodate or reinforce the activities associated with meeting those goals.

Together, these characteristics present a map of multilevel goals. The map describes how goals are addressed by organizational elements and the strength of those interactions. Let's build one such map, using Confederate, both for simplicity and to highlight contrasting sets of goals.

The starting point for these maps is the centrality of the organization's profit-making goal, which is placed at the center of both aspirational locus and organizational commitment. It's essential to understand that this means that each map is organization-specific, and comparing maps across organizations is inappropriate because placement of multilevel goals is relative to the "primary" goal of profitability.

At Confederate, there are at least four additional goals that can be assessed. The first is employee development. The second is

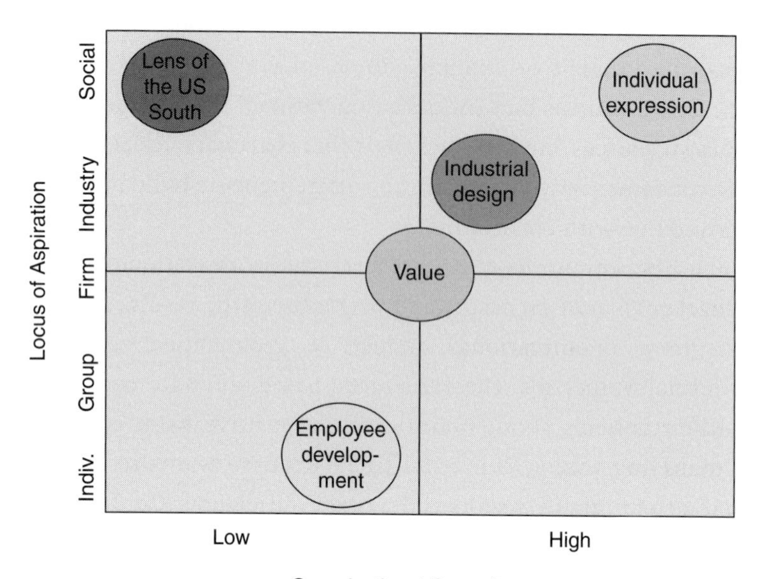

FIGURE 2.3 Multilevel goals at Confederate Motorcycles

renewing American industrial design. The third is celebrating individualization. Finally, Chambers has a preference to use the legacy of the American South, the "Confederacy," as a lens through which to understand the importance of craftsmanship in broader cultural context.

Figure 2.3 maps each of these multilevel goals for Confederate Motorcycles, based on the interviews with employees over the past two years. Keep in mind these aspects of the mapping process:

- The map is dynamic because multilevel goal characteristics change over time. For example, at the time Confederate was developing the Wraith, honoring the Confederacy was at a higher commitment level, but this became more diffused with the new Fighter.
- Some goals may extend across levels of aspirational focus. At a company like Recurve, for example, aspirational effects of ecological sustainability can be observed at nearly every level, from individual to social.

While it's helpful to make these maps reflective, the most important uses of the map will come when we look at how entrepreneurs create organizational coherence and build bridges to new opportunities. This begins with making the configuration of goals, across organizational levels, clear and explicit. Entrepreneurs and managers can use this to see the underpinnings of individual, group, and organizational behaviour. One exercise for the thoughtful manager would be to compare the map of goals against a similar map of *incentives*. While a potentially novel approach to understanding human resource management and motivation, it still reflects a mindset for achieving the expected. We want to go beyond that, to achieve the unexpected.

Balancing multilevel goals within organizations is non-trivial. Comparing goals and constraints often feels like comparing apples and oranges. What we learned from entrepreneurs is that *directly* comparing and balancing multilevel goals represents the wrong analytical process. The issue isn't whether, for example, promoting ecological sustainability is more or less important than, say organizational profitability or employee development. Focusing on these issues in the abstract generally devolves to ethical or philosophical points, heavily dependent on individual perspectives.

Every organization balances multilevel goals – they just usually aren't explicit. The first step, then, in achieving balance among multilevel goals is making those goals explicit.

The process for doing so is both simple and enlightening – just ask a sample of people at the organization what is important. We find that a measure of confidentiality is helpful – but it probably isn't necessary to bring in someone from the outside to filter the issues. Our own experience, especially with small and medium-sized businesses, suggests that there's usually 80–90 percent agreement on what's important, though rarely discussed openly. Most employees are willing to talk about those issues when asked. A short cut, though perhaps too simplistic to generate deep and consistent results, could be to explain the multilevel goal map and ask employees to identify goals either individually or in small groups. Otherwise, small group discussions provide a good basis for interpretation and goal mapping.

With a rough multilevel goal map for an organization, it's often relatively easy to identify discord.

Let's take another look at the map for Confederate Motorcycles. There are three quick lessons to take away from this particular map. The first is that it's not too crowded. There are four multilevel goals distinct from value/profits, but competing goals do not cluster in any one location. Goal clustering would suggest that the organization is struggling to clarify exactly what it wants to accomplish or how it defines success.

Confederate's multi-level goal structure is also relatively straightforward: the firm is successful if it promotes individuality and to a slightly lesser extent, industrial design. CEO Chambers encourages individual development, but as is often the case with small, specialized firms, that aspiration is secondary to keeping the firm solvent.

Finally, Chambers' efforts to honor the American South (the "Confederacy"), while evident in the architecture of the Wraith and, of course, firm branding, has faded somewhat: the P120 Fighter retains almost none of the Southern heritage, looking more like something from the aerospace industry. The release of the third generation Hellcat in late 2010 represents further movement away from rebellious individuality towards minimalism, partly reflecting the firm's push towards more efficient manufacturing. This is also reflected in the change in the firm's logo (shown in Figure 2.4) from David wielding a slingshot to an abstract design.

Formal and informal organization

As we've noted, organizational constraints and multilevel goals function within organizational structures. Thinking about business models as designs of those organizational structures helps entrepreneurs and managers leverage all of the organizational capabilities and strengths inherent in those structures without deconstructing the company into all of its functional areas and assessing those areas independently.

FIGURE 2.4 Change in Confederate logo

Organizational structures are not, however, just hierarchies or charts of subordinates and supervisors. The *formal organization* of a company incorporates these explicit relationships, but the *informal organization* often determines how a group of people interact and work. A simple definition of informal organization is the set of non-hierarchical relationships, beliefs, and interests that influence behavior. Sometimes these are very simple and common, as in a small family business where a trusted employee without ownership or executive power still controls nearly every aspect of how the business is run by right of tenure and relationship. Other characteristics may be more complex or subtle, such as the groups of individuals that eat lunch together or meet socially after work.

Business model design processes that focus on functional activities often ignore informal organization. Managers may determine, for example, that a given financial model fits with a given marketing model such as funneling profits into geographical expansion. Sometimes informal organization is addressed as a monolithic aspect of "culture" motivated by the engaged entrepreneur or executive manager. This is often perceived to be the case when an entrepreneur or charismatic manager arrives at an organization that clearly requires change. But informal organization is much more than instilling any one characteristic across an organization. Pride, quality, or cost-consciousness may be integral to success, but informal organization and multilevel goals intersect in extremely complex interactions. Imagine a global firm with an important presence in Singapore, where managers work inconvenient morning hours to maintain synchronicity with the New York office. No amount of executive-led "culture" might prevent those managers resisting expansion into China; after all, the long-term growth of that market will ultimately decrease their own influence within the organization. Unlike top-down cultural mantras, information organization, multilevel goals, and living with constraints fit together in extremely complex and interconnected relationships.

Table 2.3 highlights the tools and perspectives that innovative entrepreneurs utilize to live with constraints. Our goal is to look for the configurations and patterns that support and enable a portfolio of multilevel goals at the organization. How do we do this? We need to assess the organizational elements at firms to find ways to manage the constraints and limitations.

FRUSTRATED SYSTEMS

In 2007, ValueLabs founder and CEO, Arjun Rao, had a problem. It was, arguably, a good problem to have. ValueLabs was growing so rapidly, Rao had run out of space for his software engineers and designers.

On the other hand, this was not a trivial problem for ValueLabs, because the firm's mission centered on creating a positive and supportive workplace environment that ensured space, comfort, and even access to natural light. The company had built an award-winning facility in the high-technology nexus of Hyderabad, India. The facility emphasized workflow and worklife considerations that included recreational facilities and a canteen providing free meals to all employees. But that facility was full, and the company was projecting continued double-digit growth.

A new facility plan had been developed, and financing was obtained to help the popular, rapidly growing firm access valuable

Table 2.3 *How entrepreneurial firms live with constraints*

Tool/perspective	What it is	When it helps	When it doesn't help
Subscale effects	When local benefits or costs of resource or activity interaction within an organization outweigh higher-level benefits or costs.	Firms operating in highly specialized contexts, stable niche markets, or in markets so new or changing so quickly that customers have not stabilized into segments with clear purchase criteria.	When firms scale structures, localized conflicts may be exported to other parts of the organization. Alternately, when global contexts become relevant at the local level, subscale effects tend towards increasing costs.
Dynamic effects	When inherent change in both the environment and internal systems reduce continuity in linking organization structure to value creation.	Firms operating in markets so new or changing so quickly that the cost of experimentation is low as an option on future investment.	As opportunity shards become well-specified, firms must either routinize to lock in value creation mechanisms or bet on high-risk experimentation towards achieving the unexpected. Firms that do neither incur opportunity and/or efficiency costs.

Table 2.3 *(Cont.)*

Tool/perspective	What it is	When it helps	When it doesn't help
Complexity	Even small, simple systems may generate unpredictable outcomes.	Firms with clear priority maps and flexible structures may leverage complexity and take advantage of unexpected value.	When firms neither effectively reduce complexity effects through structuring nor use complexity to explore value creation processes.
Multilevel goals	Simultaneous goal sets across the organization of varying priority and locus of aspiration.	Firms may shift between goal focus based on opportunity exploration without losing coherence or continuity.	Firms with dense goal clusters may be unable to distinguish between goal priorities; firms that fail to acknowledge goal priority shifting suffer dissonance.
Formal and informal organization	The complete set of organization design structures that bound and direct resources and activities.	Firms that leverage structures in developing coherent sets of elements may overcome inherent constraints that otherwise can't be reconciled.	Inconsistent formal and informal structures exacerbate existing conflicts within organization systems.

land. But there was a hitch: final government approval was being held up on technical details. It was becoming clear to Rao that, as happens in India and elsewhere, pushing the project forward would likely involve ethical compromises, most likely in the form of payoffs to key third parties. Rao had built his firm on a foundation of integrity – he'd put "value" right into the company's name. Starting over would delay available space for at least two years, and rental costs in comparable space in Hyderabad would be prohibitively expensive.

In the end, Rao saw through the frustrated system: ValueLabs turned down the offer of land for the much needed building. The premise and narrative of how the company was built and shaped did not quite fit the opportunity at hand. Compromising the integrity of the firm would create more frustration internally than living with current constraints.

Highly effective organizations ...

No matter how skilled the entrepreneur or manager, no matter how motivated the employees, no matter how well-crafted the business model, no firm is perfect. Our idealization of highly effective organizations like Southwest, Ikea, Apple, Google, Wal-Mart, Nordstroms, Tata, and Honda, are built on oversimplifications, selective vision, and post-hoc rationalization. It only takes a little history and hindsight to uncover flaws in previously perfect strategies and business models. Enron was a poster-child for business model innovation – until suddenly it was a poster-child for fraudulent accounting practices. Honda's entry into the United States motorcycle market in the 1970s was hailed as brilliant planning, market segmentation, and implementation, until it turned out that it happened partly by accident and partly by smart people making short-term decisions with no long-term strategy in mind at all.[20]

Take Southwest Airlines again, the only consistently profitable airline in the United States for more than three decades. That success is attributed primarily to the firm's unrelenting dedication to keeping costs low. And yet, Southwest is at the industry average for fleet age among the six major US carriers, 12.7 years vs. 12.9 years,[21] and has a newer fleet than United, American, and Delta. Is this a perfect fit with the lowest-cost strategy? This isn't to say that Southwest doesn't have a good strategic vision, or even that it isn't implementing a great strategy, only that perfect strategic complementarity is probably impossible. Tradeoffs, physical reality, black-swan events like 9/11, and sociopolitical contexts at organizations get in the way. Again, we aren't suggesting that Southwest's success is not linked to its low-cost strategy, just that the strategy isn't representative of perfect complementarity across all the firm's resources and activities. The perception of perfection may be convincing, but it is an illusion.

Let's consider some of the entrepreneurial firms from our study. Cellular Dynamics, profiled more extensively in Chapter 3, may be the world leader in stem cell technology. As of Spring 2011, it was the first and only company to be providing a stem cell-based assay for drug testing that is not based on human embryonic stem cells. This is an organization operating at the absolute leading edge of life sciences research. The launch of CDI's iCell™ cardiomyocytes brings stem cell-based technology into full commercial scope. This is a highly effective organization.

Return Path is another example. No other company in the world comes close to matching its 2-billion email inbox database for email deliverability. Recurve has piled up the accolades, including *Inc.* 5000 status with triple-digit growth and Fast Company innovation awards. We have already noted effectiveness at Confederate Motorcycles.

But none of these are perfect organizations.

With frustrated systems

Return Path presents numerous non-mutually enhancing interactions. These include conflicts associated with geographic dispersion, divergent goals between the receiver-focused and sender-focused development groups, and the inherent conflicts between the employee development program and near- and mid-term product launch requirements.

Confederate's conflicting elements are even more dramatic, starting with the firm's publicly listed financing scheme counter-indicated by the low-volume sell-make manufacturing model and social/philosophical multilevel goals. Most analytical finance tools would strongly argue against public listing.

Recurve operates an inherently frustrated system. The construction and software businesses have dramatically different systems across the organizational spectrum: compensation, hierarchy, job types, business model components, risk-reward profiles, asset specificity, and so on. Amazingly, the businesses share office space, senior management, and information systems.

The bottom line is that most systems are imperfect. In particular, most systems are "frustrated": the necessary interactions between people, groups, goals, processes, and narratives preclude the possibility of creating perfectly reinforcing relationships.

Frustrated systems are not, however, necessarily bad or even competitively disadvantageous, at least for entrepreneurial firms. And, sometimes, they are simply unavoidable. In order to fully appreciate these facts, we'll look at a simple example to clarify how frustrated systems can be identified and typed. Then we'll examine the structural mechanisms entrepreneurs utilize to reduce frustration. In Chapter 3 we'll show that at entrepreneurial firms, especially interesting entrepreneurial firms, the organizational structuring can actually be modeled based on a very different strategic framework from complementarity.

REDUCING FRUSTRATION

This brings us to a critical aspect of the entrepreneurial manager's role leading organizations to achieve the unexpected. The design of highly effective organizations achieving the unexpected *minimizes frustrations rather than targets perfection.* This is a completely different approach from traditional strategic planning frameworks.

There are three relatively simple rules to make this possible:

1) Violate weak constraints rather than reconfigure conflicting elements
2) create subgroups of similar elements
3) strive for locally stable systems.

To see this type of organizational management in real-time, spend an hour watching a well-regarded kindergarten teacher manage a classroom of highly idiosyncratic children. Perhaps this seems trite or nostalgic, but the lesson isn't what the children learn, but the mechanisms a competent, entrepreneurial instructor uses. Consider the following: kindergartners won't always agree, and their emotions will dominate logic. Solution: instructors accept that not every child can be happy all the time, and weak constraints will have to be broken to serve the benefit of the group. The children cannot all sit next to the teacher or feed the hamster. Young children develop emotionally and physically at different rates – randomized group play carries important social benefits but runs the risk of generating significant conflicts. Savvy teachers group children who work and learn well together, and change those groups as capabilities and preferences rapidly change, sometimes on a daily basis. Finally, managing a kindergarten classroom isn't an exercise in monolithic consistency: teachers who demand total conformity are likely to engender chaos and opposition, especially once the novelty of conformity wears off. Truly effective kindergarten instructors create pockets of stability, often reinforced structurally by "stations" throughout the classroom. These support the creation of locally stable systems, even though the children at those stations may change minute to minute.

In a more serious vein, consider a government finance committee that incorporates members of different parties, especially in a multi-party parliament such as the UK, France, or Israel. These are clearly frustrated systems, and yet somehow progress, though of varying quality, is accomplished. Even the American Congress manages to pass reasonably useful legislation on a semi-regular basis.

Let's take these lessons and address them in apparently more complicated and less well-understood human systems: companies.

Violating weak constraints

In mathematics, constraints are the limitations or rules that determine how network nodes interact. We can think of constraints at a variety of levels, but for our simplified modeling purposes we're going to assume that all constraints can be described as the relationships between the nodes. Again, for simplicity, we'll assume that these relationships have two characteristics: strength and polarity. The polarity of the relationship determines whether it is reinforcing or conflicting, and the strength of the relationship is simply how powerful that effect is.

The first lesson for managing frustrated systems is to violate weak constraints rather than reconfigure the system. All too often, managers strive to accommodate every person, group, and relationship. They sometimes use informal organization, attempting to generate a holistic sense of community in which everyone supports everyone else in all operations. Alternately, they attempt to reconfigure the entire system to somehow eliminate all conflicts. Our observations of entrepreneurial firms suggest that these efforts are well-intentioned but often misguided. Organizational elements are sometimes incompatible enough to defy most reconciliation efforts. This may be seen as a "lesser of two evils" argument or even "better the devil you know," but it really reflects a larger function of system-based relationships. In addition, it is sometimes those conflicts that lead to the unexpected.

In Cellular Dynamics Inc. (CDI), maintaining operations under one roof meant violating weak constraints associated with interaction between the research and development groups. The tensions associated with these problematic interactions occurred primarily at the middle management levels. Project and product managers ran into scheduling and resource utilization conflicts in which no clear priorities could be clarified, because the separate organizations had distinct timeframes for activities and unspecified shared access to equipment and people. Staff were frustrated by these conflicts but expressed the opinion that resource allocations problems would get resolved by their direct managers. Senior executives were likewise aware, but had delegated the problem to middle managers without explicit prioritizations. In fact, our interviews with middle managers revealed a diluted "fog of war" mentality. Middle managers expressed an interest in more clear prioritizations, but felt that muddling through was a viable near-term option.

Weak constraints often look like strong constraints from a strategic perspective. At CDI, the distinction between therapeutic and tools development activities look like completely separate business models: long-term vs. short-term, high-investment vs. low-investment, high P/E vs. low P/E, research intensive vs. development intensive, and so on. Most traditional strategic analysis would segregate the value chains and focus on separate financing and resource development. In fact, the system of uncertain priorities and shared "pain" may have been directly responsible for one of the firm's major breakthroughs in manufacturing scale-up. Although a specific work team had been established to develop protocols for efficiently scaling cell culturing processes, those efforts hadn't been successful. In the end another work group developed an alternate solution, initially very controversial within the organization, but ultimately proving to be the better solution.

At Confederate, a similar weak constraint links the firm's legacy parts group with active manufacturing sourcing. Confederate's lengthy history includes a difficult period during the recession of 2000–2001 when the Board fired Chambers and attempted to ramp up manufacturing of the firm's iconic Hellcat cycle by cutting costs. For a variety of reasons, the effort failed and Chambers ultimately bought the company out of bankruptcy. In 2005, Hurricane Katrina destroyed nearly all the company's records, including design and specification information for the Hellcats. The Hellcats manufactured during the 2000–2001 period are more prone to mechanical problems; but the company simply doesn't have the records or specifications for the various parts of the bikes. When Confederate customers who paid $35,000 for those Hellcats need a replacement part, Confederate often has to design and build replacement parts by hand based on customer photos rather than CAD drawings.

While a relatively profitable sub-business on a loaded-cost basis, the activity detracts from the firm's ongoing focus on new designs and more efficient craftsman-based manufacturing. But it is a weak constraint compared to the link between the firm's resource-intensive brand and the legacy parts group. Prior owners are both highest-potential targets for new purchases and strong influencers in new client acquisition. The organizational effort associated with fully reconfiguring activities would likely either fail or impose higher costs and conflicts on the system.

Create subgroups of similar elements

The second lesson is to create subgroups that comprise similar elements. At CDI, there are three primary subgroups:

1) Therapeutics, research orientation, blood research
2) tools development orientation, materials development
3) automation skills, stem cell culture skills.

The benefits of these groupings are relatively self-evident. Coordination costs are reduced, managerial attention is focused on operational efficiency and outcomes rather than inter-group conflicts. The challenge, however, is ascertaining which subgroups have the right characteristics to function well together.

It's interesting to note that in the conceptual model we'll develop in Chapter 3, subgroups 1 and 3 fit into a larger subgroup. This is, in fact, in contrast to the legal organization at CDI prior to the merger, where subgroups 1 and 2 functioned in one entity and subgroup 3 in the other. There are a number of lessons to take away from this observation. First, it's quite possible that CDI's organizational structure was suboptimal given the complex interrelationships among business model elements, despite being organized along apparently logical functional lines. It's important to remember that most frustrated systems do not have a single optimal configuration, but more likely present multiple suboptimal but stable configurations.

Our analysis of IBM's Global CEO data, which is discussed in Chapter 4, shows that there are patterns for how firms, including very large firms, adjust structures while addressing new opportunities. But rather than reconfiguring systems, they tend to simplify, which appears to facilitate managerial attention towards opportunity identification. These firms likely do pass through areas of lower efficiency, in particular associated with the exploitation of current opportunities. But bringing conflicting elements together by simplifying structures serves a larger goal – preparing the organization to identify and exploit nascent markets previously inaccessible.

We are not advocating that highly conflicting elements be forced into proximity. One memorable consulting project from our own experience springs to mind. An entrepreneur running a highly specialized truck and tractor engine repair business in the US Midwest wanted to grow the business by adding a rapid automobile oil change service for commuters. Much of the equipment and supplies would overlap, and service bay space could be managed between the services to increase capacity utilization. The obvious problem, however, was that the truck and tractor technicians had custom skills based on decades of experience with a variety of complex tractor and truck engine types, while the "grease monkeys" who would perform oil changes were low-skill, transient employees usually 20–30 years younger. But they'd have to share tools, equipment, and even a lunchroom. The economic theory is plausible, but the probability of a frustrated system even more so. To the extent that the organization could be designed to separate the subgroups, the system might function, but facility and supplies overlaps didn't bode well for the effectiveness of the organization as a whole.

Create locally stable systems

In the kindergarten class, "stations" of activities provide a physical structure that often functions effectively regardless of the characteristics of the children at the stations. Over time, of course, we might find that some children tend to gravitate to certain stations. Regardless, what we see at any given station on any given day are locally stable systems – a microcosm of how subgroups function within any organization.

Strategic theory has generated entire service industries focused on finding the optimal way to structure organizations. These derive from research suggesting that certain organizational configurations are always optimal for certain technologies or strategies. But the data supporting these arguments is weak at best. Efforts to generate universal templates for optimizing structures at organizations go in and out of fashion. Recent research suggests that organizations may even be structured to take advantage of non-objective capabilities such as intuition.[22]

Creating a generic, template structure approaches organizational design as comparable to machine design subject to specific physical laws and engineering approximations. Elements of the system are presumed to be well-defined and potentially replaceable, rather than irregular and idiosyncratic. Redesigning a car door to improve the perception of quality vis-à-vis heaviness, the feel of the handle, the sound when it closes, is an entirely different problem from redesigning an organization to produce higher quality car doors.

Applying universal templates necessarily means ignoring the elements that make subgroups locally stable. At Savage, for example, the art design group functioned on a completely different wavelength than the engineering group, often at completely different hours. These kinds of functional differences are common and well-established, but not as obvious in the case of cross-functional teams and multi-capability teams in larger organizations. And it is possible to support locally stable but conflicting functions in close proximity with the right structures and control mechanisms. When CEO Pratap Mukherjee led the massive transition at Recurve from a construction company to a software business that supports green construction, he decided he needed the in-house energy auditing and construction business to serve as a market-facing source of customer information and product testing. But maintaining a construction business inside a venture-financed software business is no simple task, especially when the construction staff are fully aware that the future of the company is in the hands of the software programmers.

The management team at Recurve has worked to carefully balance the relationship between the construction business and the software development effort. Company-wide activities maintain the link between the groups, while profitability for the construction business has been segregated from the organization as a whole and assigned to a manager with direct experience managing construction projects and companies. It's not an obvious balance – the metrics for success of a construction company look nothing like those for a software company, and every aspect of data collection, motivation, and monitoring must be adapted to the distinct group characteristics.

But so far the system is working – Fast Company named Recurve the third most innovative company in the Energy Sector in the US.[23]

Opportunities are not perfect. Neither are organizations.

We've seen, however, that a traditional conceptualization of strategy as perfect internal consistency is neither realistic for entrepreneurial companies, nor a necessary condition for success. Every organization lives with constraints; highly effective organizations present frustrated systems in which perfect internal consistency and mutual reinforcement simply are not possible. Table 2.4 summarizes when mechanisms for reducing system frustration work in entrepreneurial organizations.

Table 2.4 *Reducing frustration at entrepreneurial firms*

Mechanism	Works when ...	Fails when ...
Violate weak constraints	• Multi-level goals counteract internal conflicts • Formal and informal systems combine to provide alternate support mechanisms • Internal conflict serves to highlight developing opportunities	• Conflicts have unseen ripple effects across the organization • Weak constraints increase in importance as new opportunities arise • Internal conflict is purely logistical or interpersonal
Create subgroups	• Organizations have relatively well-defined capabilities and connections that can be segregated without significantly disrupting formal structures • Conflicts between groups, even strong conflicts, can be balanced or mitigated with informal structure • Subgroups can be linked to high-potential, nascent opportunities	• Conflicts between subgroups feed systemic perceptions of unfair management practice • Subgroups become too internally focused to identify nascent opportunities

Table 2.4 *(cont.)*

Mechanism	Works when ...	Fails when ...
Create locally stable systems	• Competitive pressures are relatively low because opportunities are ill-formed or poorly understood • Competitive pressures are so high or industry dynamics so unstable that change is inevitable and frequent • Stability facilitates attention to nascent opportunities	• Competitive environments are intense but stable, requiring focus on incremental product and process innovation rather than business model innovation • Stability encourages focus on extant opportunities rather than changing market conditions

There are better ways to approach the entrepreneurial process than the relentless pursuit of illusory perfection. Entrepreneurs pursuing truly novel opportunities should be aware that organizations can function effectively with conflicting elements, especially at subscale levels. In addition, managing complexity can be achieved by balancing multilevel goals rather than enforcing conformity or attempting to satisfy every desirable organizational relationship.

Companies like Cellular Dynamics, Recurve, Return Path, Confederate, and ValueLabs have learned these lessons already, challenging common sense rules about competitive strategy. They have discovered after years of struggle that there is something much more important than strategic positioning. It is the key to the effectiveness of imperfect organizations that function in an imperfect world. And, although the entire theory of imperfect organizations seems to suggest that complexity would render any analytical effort useless, there is a surprise hidden in the very nature of that key.

The key is coherence. The surprise is that it can be modeled. In the next chapter, we will model coherence to show how managers remodel organizations.

MINIMIZING FRUSTRATION AT BROADJAM

There is no doubt that Broadjam's survival and recent success is a testament to the dedication and resilience of its founder and employees, the patience of its investors, and the importance of networking. At least as important as all of these, however, is the narrative of Broadjam's business model, which has been mixed and remixed to reduce frustration within the organizational system without yielding the core vision.

The recent economic crisis has not been kind to most small firms, especially in service businesses that rely on revenues from consumer disposable income and medium-sized companies as clients. Broadjam derives a majority of its sales from subscriber musician hosting fees and advertisements on its website targeting those musicians. Yet, Broadjam is seeing web traffic and revenues increase.

Some of this is likely due to counter-cyclical elements: media buyers looking for lower-cost content and independent musicians hoping to augment income via a beloved hobby. But the base of the story is more important than (even turbulent) current trends. Broadjam developed a coherent narrative that enabled a flexible approach to business model evolution.

When Roy Elkins left Sonic Foundry, he'd already "made it." He had pursued work in the industry he loved and combined natural networking talents with good timing. Elkins had been an amateur musician who started at the ground floor, selling musical instruments retail. His technical proficiency, networking skills, and passion for music led to senior marketing and sales work for Ensoniq and then, in a significant leap of faith, to VP of sales and marketing for Sonic Foundry, a media software and tools company. He helped grow the company to an IPO that fitted the exuberance of the late 1990s: following a relatively early IPO (revenues of only $15 million) the stock price soared to $60/share in early 2000 giving the firm a market capitalization exceeding $1 billion. Elkins left Sonic at its peak and was able to cash out some of the stock before the bubble burst; Sonic's stock fell below $2/share and the firm entered a lengthy period of slow organic growth.

Elkins had seen the growing importance of technology to musicians – Sonic Foundry's tools enabled artists to use a PC to manipulate media with the power of a professional studio. But Elkins saw music as primarily a social and cultural phenomenon, and it was a different technological marvel that fed his true enthusiasm for empowering musicians: the Internet.

When he founded Broadjam in Madison, Wisconsin, online access to music files was in a nascent stage. Mp3.com was offering free downloads in 1997, and the first mp3 players had become available for mass retail in 1998. Mp3.com's IPO raised $370 million in 1999, and the company traded at a $3.5 billion valuation. The market seemed to be tilting towards non-traditional distribution processes that would favor consumers and artists over the recording labels. By

2000, the entire industry was in turmoil. The major record labels were up in arms about copyright violations. Unlike cassette tape copies, mp3 files were effectively instant and costless for both the sharer and recipient. Perhaps the record companies had seen the writing on the wall – compact disc sales would peak in the US (and worldwide) in 2000 and drop 5–10 percent each succeeding year.[24]

Critics of the industry argued that music should be freely shared, like books, while artists, recording studios, and record labels found themselves in the unenviable, but apparently unavoidable position of suing their customers and distribution channels.

Elkins knew that the music industry is one of the most idiosyncratic businesses; success is based on a strange and unpredictable combination of talent, connections, circumstances, and luck. If Broadjam could create a critical mass of independent musicians, it could potentially sign musicians to a publishing contract and become an established piece of the music industry money machine. Hit songs can generate millions of dollars for songwriters, artists, and promoters – and music rights are now valid for 95 years. One big hit could generate massive returns to Broadjam and its investors.

Elkins' vision was to help independent musicians commercialize their art by leveraging a shared infrastructure of technology, connections, know-how, and savvy. The initial business model presented the company as a collection point for independent artist's work, distributing content to the burgeoning community of online musical distribution nodes like mp3.com. The artists would be "discovered" on those sites and Broadjam would benefit from the downstream work of arms-length relationships. Customers (artists) would pay Broadjam to maintain the appropriate file type and content management information so that their songs would get appropriately labeled (by genre and "hook" – the key lyric or musical riff that makes the song immediately memorable) and distributed to the sites most likely to generate exposure for the artist. Elkins raised local angel money based on this model, the diagram of which had become the talking point for nearly every discussion about the company.

By 2001, the company was struggling. Even as the number of online music downloads accelerated exponentially and compact disc sales faltered, the recording industry retrenched behind the legal protections of copyright law. In conjunction with major artists, the industry sued the online distribution sites as well as the consumers using those sites to obtain free music. Broadjam found itself in unknown territory: online music access was inevitable, but the record labels were apparently willing to burn their near-term profits in litigation to preserve long-term rights. In the meantime, the dot-com bubble had collapsed and venture funding across all technology sectors had dried up. Elkins used his personal capital reserves to prop up the company and limit his own salary needs during lean years.

Broadjam's entire business model was at risk: the firm's resource structure required growth to reward investors and retain key employees, the transactive structure required downstream distributors to promote artists, and without a track record of successful hits, the company would struggle to sign new artists or capture the value of its efforts. Businesses with flawed business models do not survive.

But Broadjam did survive. Some factors in Broadjam's survival include Elkins' determination and drive, the fact that he had financial resources to fall back on during lean times, the patience and financial resources of the Broadjam investors, and the culture of enthusiasm and trust among the employees of the company. But none of these elements are sufficient or even absolutely necessary to the success of a business. In the end, Broadjam survived because *Elkins allowed the narrative of the organization to support, rather than reject, changes in the business model* through multiple environmental changes and internal innovations.

Three events have marked the critical junctures for the viability of the Broadjam business model. The first was the collapse of free/illegal online music downloads in the wake of the dot-com bust. Although various incarnations of these models still exist, none of the lead players or seemingly dominant technology platforms survived the legal shakeout in large-scale viable form. Broadjam's first transactive structure presumed that some variant of an integrated and widely available music distribution platform would be available. This was true in 2000 and 2001, but not in 2002, 2003, and 2004.

The collapse of venture investing, interest in so-called dot-com companies, and chaos in the music distribution industry was a perfect storm, preventing Broadjam from raising funds, signing independent artists, or generating revenues from any type of distribution mechanism. Elkins and the entire staff at Broadjam were confident that the technology they had built was licensable and scalable. The Broadjam business model was based on the concept of building a database of music and related metadata, but without a transactive structure linked to monetization, the resource structure of the business model effectively existed in a vacuum. The company was bleeding cash.

For Broadjam, survival depended on a second, somewhat unexpected opportunity linked to its primary opportunity. Music directors for TV, film, and advertising were once at the mercy of the music publishing companies. There was no viable way for them to access, review, and select musical content for their media needs without the direct involvement of the labels and studios. The music supervisor of a film, TV show, or advertising firm would contact a music publisher and request samples of songs in a particular genre, perhaps referencing certain words or themes. The publisher would then review *physical printouts* of the songs under license and, based primarily on memory and individual musical sense, submit a selection of songs for consideration along with the license fee requirements for each option. The publisher was the bottleneck. A premium was placed on recognized artists, because a positive response from consumers was effectively guaranteed, though at a very high price. The key experts at the publishers were human repositories of data, stored in highly tacit structures of associations, recollections, and subjective evaluation of merit.

Elkins had connections to a variety of industry execs. He recalled a meeting with a senior executive at one of the large record labels. In casual conversation, he'd asked the executive a question about the label's music catalog, and had been surprised to see the executive extract an enormous printout from a locked cabinet. The printout was on continuous form paper, folded accordion style with holes punched along the sides, clearly generated by a line printer. Even for the late 1990s this was clearly a relic of old technologies. Given that the label's catalog of music included millions of songs, retaining the catalog list in printed, unfiled form was a significant surprise.

Elkins and the Broadjam investors knew that focus on long-term value creation is essential to the success of technology-based ventures. But without revenue, Broadjam's technology infrastructure would be rendered valueless by the firm's inability to monetize it. The team made a difficult decision: they took consulting contracts to generate cash. Elkins decided not to take any consulting business that was outside of the core music industry competency, or any projects that would not directly help build the key capabilities of the firm's technology base.

He contacted the same music publisher and described how Broadjam's proprietary music data management infrastructure could be applied to the firm's archive music catalog. In a testament to the value of that platform, Elkins' relationship-building skills, and the integrity of the venture, Broadjam continues to support that contract eight years later. Other related projects followed, and Elkins righted the sinking ship. The company remained focused on two key aspects of the venture narrative. First, that the ability to manage music content and the metadata associated with that content required both technological and artistic savvy. Second, that the value of music content management could be monetized once a clear transactive structure could be legitimized.

The third evolutionary phase was the legitimization of online music distribution. By 2004, numerous models were in place, including Broadjam's own online song purchase system. But standardization was triggered by Apple's launch of the iTunes Store. For the music industry, this event represented a watershed in the transition to electronic media: iTunes validated the legal music download space by offering 99 cent downloads from nearly all the major record labels. Songs were encoded with digital rights management (DRM), though users of iTunes could still burn CDs that could be illegally shared. The success of the iTunes Store is hard to overstate – as of January 2009 iTunes Store users had purchased 6 billion downloads, accounting for approximately 70 percent of all legal downloads worldwide,[25] and in 2010, the 10 billionth song was downloaded. By comparison, music CD sales peaked at $15 billion in 1999. By 2010, sales of music CDs had fallen to $6 billion, while iTunes had grown to a $5 billion annual revenue stream. In 2011, iTunes surpassed CDs as the dominant distribution channel, though this was eclipsed by the even more dramatic success of the iTunes App Store, which generated more than a billion downloads in the first year, and 10 billion in three years. But back in 2004, the bottom line was that music was moving to legal electronic media, and the record labels had found a way to move with the market rather than against the market, even if the solution was implemented by an outsider.

There is a significant irony to this story worth noting. In the early days of the Internet, the collapse in telecommunications pricing structures led many to suggest that the long-term value in ICT would have to be in content. The Time Warner-AOL merger was, perhaps, the most significant transaction based on this idea. There remains much truth to the argument, but it is useful to note that of all the players in the music industry, it seems likely that it is neither the physical infrastructure nor the content providers that have generated the most value. It is Apple. Apple neither produces the music nor provides the data transport. But Apple found a way to port its consumer experience business model into an industry that had no viable transactive structure.

For Broadjam, living payroll-to-payroll, the impact was not immediately obvious. Elkins looks back on iTunes as the seminal event in the long-term success of Broadjam, but doesn't recall thinking about it at the time as anything more than another venture into legal downloads. But for Broadjam, along with many other organizations, the success of iTunes would epitomize the transition of an entire industry, and especially the entry-level space, where new musicians get their first big break. The advent of electronic music content, and more importantly, the ability to track metadata associated with music content, began to shift power to the music supervisor. He could request and review significantly more content based on the metadata descriptions, including content from independent artists and other catalogs of music not controlled by the major publishers or labels.

As the new world of music content management evolved, it became clear Broadjam had precisely what music supervisors wanted: a sortable, searchable database of music from independent artists. And, by extension, Broadjam had what independent musicians wanted: access to music supervisors actively seeking unsigned music. The transactive structure of the business model was validated, nearly a decade after Elkins first came up with the idea.

The Broadjam business model in 2000 was at least *four years too early.* The underlying core technology for storing, tracking, and accessing music electronically, including the metadata that music supervisors need, was already in development and would be perfected as the company grew. The online interfaces and processes to help music supervisors access unsigned, independent musicians were also in development, and the independent musicians needed to know that there would be viable mechanisms to propagate their music online. All of this was finally in place in 2004, though Broadjam wasn't in a position to capitalize on it. For Broadjam, the business model that was too early in 2000, salvaged in 2002 for survival purposes, and limited from 2004 to 2007 by tight resources and slowly developing relationships with music supervisors, finally grew into its own in 2008. Broadjam began providing placements as a pay-to-play service to independent musicians as part of their membership agreement. The possibility of realizing the most lucrative aspect of Elkins' original business plan, serving as a licensing clearinghouse for unsigned music, remains untested while the company manages growth stemming from the business model change in 2008. But the vision of serving independent artists remains intact and vibrant.

The company has dramatically increased its customer base, due in part to the firm's success in placing artists into hundreds of commercial opportunities in advertising, TV, and film. At the request of those same artist-customers, Broadjam launched a free song download service – not unlike shareware software download systems – giving artists an opportunity to share their music in the hopes that fans will then buy more. The company sponsors and supports a variety of worldwide music contests, creating ever more outlets and press for its artist base. One of the most fascinating elements of the Broadjam story is that the company hosts artists from 190 different countries – almost every country on earth.

Will Broadjam be a top ten hit? Probably not. In terms of commercial success, Broadjam is unlikely to rival the big labels like Warner Music or Universal/Vivendi, now that those organizations have effected migration to the digital age. The apex of the music industry still requires enormous investments in public relations and Hollywood culture. Despite experiments like Radiohead's *In Rainbows* which was offered online for whatever price customers wanted to pay, most major artists remain tightly linked to the big labels. But the Internet continues to drive change in the music industry, and the Broadjam business model now creates and captures value previously inaccessible to a significant community of "hobby" and "wannabe" musicians worldwide.

NOTES

1 Napster was resurrected in 2006 by negotiating new contracts for legal music distribution, but has had little impact on the industry. Its revenues are dwarfed by iTunes.

2 "Paul van Riper's big victory," in Gladwell, M. 2005. *Blink*. New York: Little Brown & Company.
3 Shane, S. 2000. Prior knowledge and the discovery of entrepreneurial opportunities. *Organization Science*, 11: 448–469.
4 www.prweb.com/releases/2011/9/prweb8777406.htm (accessed September 15, 2011).
5 Source: www.en.wikipedia.org/wiki/Peter_Molyneux (accessed December 12, 2010).
6 Sources include interviews with Savage management and online documentation including www.en.wikipedia.org/wiki/Game_development (accessed December 12, 2010).
7 Anderson, C. 2006. *The Long Tail: Why the Future of Business Is Selling Less of More*. New York: Hyperion.
8 The recent success of small gaming "apps" available on mobile and browser-based operating systems could be seen as a reversion or rebuttal. The *Angry Birds* game, for example, has generated more than 200 million downloads on a variety of platforms. In fact, however, apps appear to represent an entirely new market, but do not, as yet, appear to compete directly with the video gaming industry, which generally has seen strong sales, taking into account the significant economic downturn.
9 See, for example, "Why video game developer acquisitions scare me" by Don Reisinger for cnet.comm, www.news.cnet.com/8301-13506_3-10188950-17.html (accessed December 12, 2010).
10 "Dean Kamen's 'Luke Arm' prosthesis readies for clinical trials" by Sarah Adee in IEEE online, February 2008. www.spectrum.ieee.org/biomedical/bionics/dean-kamens-luke-arm-prosthesis-readies-for-clinical-trials/2 (accessed December 1, 2010).
11 Source: www.jaipurfoot.org/01_visitorviews_presscomments.asp (accessed November 15, 2010).
12 Source: www.time.com/time/specials/packages/article/0,28804,1934027_1934003_1933963,00.html (accessed November 15, 2010).
13 Wernerfelt, B. 1984. A resource-based view of the firm. *Strategic Management Journal*, 5: 171–180.
14 Milgrom, P. and Roberts, J. 1990. The economics of modern manufacturing: technology, strategy and organization. *American Economic Review*, 80(3): 511–528.

15 "Airline Unions seek a share of the industry gains," *New York Times*, October 27, 2010, www.nytimes.com/2010/10/28/business/28labor.html?adxnnl=1&adxnnlx=1296574366-DODW2t498CMyIvaIWlnNqA (accessed November 1, 2010).
16 "Ex-Southwest Airlines CEO James Parker shares lessons in book," Terry Maxon for the *Dallas Morning News*, January 20, 2008.
17 Source: www.blog.cloudharmony.com/2010/02/cloud-speed-test-results.html (accessed July 1, 2010).
18 Stinchcombe, A. L. 1965. Social structure and organizations. In: March, J. G. (ed.), *Handbook of Organizations*. Chicago: Rand McNally & Company, pp. 142–193.
19 "Speed demon" by Phil Patton, *ID* Magazine, June 2005 (accessed via Confederate.com website on July 1, 2010).
20 Pascale, R. T. 1984. Perspectives on strategy: the real story behind Honda's success. *California Management Review*, 26: 47–72.
21 Calculated from 2009 data from www.airfleets.net.
22 George, G. and Bock, A. J. 2010. The role of structured intuition and entrepreneurial opportunities. In: Phillips N., Griffiths, D., and Sewell, G. (eds), *Technology and Organization: Essays in Honour of Joan Woodward (Research in the Sociology of Organizations, Volume 29)*. Oxford: Emerald, pp. 277–285.
23 Source: www.fastcompany.com/mic/2010/industry/most-innovative-energy-companies (accessed November 1, 2010).
24 RIAA 2007 Year-End Shipment Statistics, www.RIAA.com (accessed July 1, 2010).
25 Phil Schiller's keynote address at MacWorld Expo 2009 as reported by *MacDailyNews*, www.macdailynews.com/index.php/weblog/comments/19613/ (accessed November 1, 2010).

3 Remodel for coherence

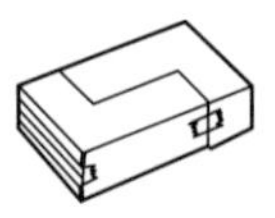

SHAPING THE FRONTIERS OF HEALTH SCIENCE

In a modest building in the University Research Park in Madison, Wisconsin, extraordinary science guides a company towards some very implausible opportunities. Despite a full-time employee count under 100, Cellular Dynamics International (CDI) boasts the highest concentration of stem cell knowledge in the world – more than the transnational pharmaceutical companies, more than the leading-edge universities studying stem cells.

As of early 2011, there are still no approved drugs, therapies, or treatments of any kind based on stem cell technology. Although originally identified in the 1960s, human stem cells were first derived by Dr. James Thomson at the University of Wisconsin-Madison in 1998. The technology has been awash in political, legal, religious, ethical, and scientific controversy ever since. The US federal government has restricted how federal scientific funds are used for stem cell research, while states like California and Wisconsin have made public funds available for both educational and corporate research activities. The global picture is even more chaotic, in part because the foundational patents owned by the University of Wisconsin-Madison do not cover use outside the US.

The development of induced pluripotent stem cells (iPS) may provide a less controversial source of non-embryonic stem cells, but the debate on stem cell research and use is unlikely to fade away anytime soon.[1] Venture funding for stem cell companies peaked at approximately $250 million in 2000 and subsequently fell to $50 million by 2004. Funding has rapidly expanded again in recent years, with dozens of eight-figure funding rounds in 2009 and 2010, despite the economic recession. There are more than 20 publicly traded firms and hundreds of privately funded firms focused exclusively on stem cell research. The total annual investment in stem cell research is almost impossible to estimate, in part because extensive research efforts are ongoing at universities, government laboratories, and companies in places where limited reporting is available.

The lure of stem cell technology is powerful. The potential to heal or regenerate dead or dying cells in a patient as needed, repairing or replacing a faulty or diseased heart, liver, lung, or kidney, presents a Star Trek vision of future health care. Rumors of experimental stem cell treatments in places like China have turned out to be true, luring patients with otherwise untreatable conditions despite uncertainty as to whether or even why such treatments work.[2]

In this implausible industry, CDI is, in many ways, even more of an oddity. Despite the founding role of Thomson, CDI has obtained venture funding regionally, without support from any of the big East or West Coast venture capital firms, despite the fact that Wisconsin has one of the smallest venture capital sectors in America. The company remains based in Madison, population 200,000, without direct flights to San Francisco or Boston. The firm's first product is living, *beating* heart cells in a petri dish for safety testing of potential therapeutics. And, just to make things interesting, CDI used to be three companies with three different business models until they were merged in 2009, defying traditional strategy principles that separate value chains optimize competitive positioning and leverage core competencies.

INSIGHT 3: REMODEL THE ORGANIZATION FOR COHERENCE

The third insight to achieving the unexpected is to remodel the organization for coherence. Entrepreneurs should understand how firm structures are linked to plausible narratives at the organization. This explains one of the core distinctions between traditional strategy and entrepreneurial action pursuing unusual or novel opportunities. Whereas strategic managers are focused on optimizing organizational systems to position the firm effectively in the focus industry, entrepreneurs link organizational structures to the process of making sense of opportunities.

This argument may seem strange at first glance. In fact, many entrepreneurs understand it intuitively. In addition, it builds directly from key research that approaches entrepreneurship as a fundamentally creative endeavor. This perspective frames entrepreneurship as an emergent process that develops and deploys novel resources and constructs entrepreneurial meaning as a social process within organizations.[3]

To develop and describe this insight we'll examine CDI in detail through this fundamentally entrepreneurial lens. We'll also refer to other firms, both from our study and in broader context, to provide more examples for how entrepreneurs participate in nothing less than shaping how all of us understand innovation and change. First, we'll follow up on the concept of imperfect opportunities and living with constraints to appreciate why strategic "fit" may simply not be relevant for tomorrow's entrepreneurs. This also provides more perspective on multilevel goals and how the narratives that structure decision-making at innovative firms change and evolve. Finally, and perhaps most important, we'll lay the groundwork for an entrepreneurial model to achieve the unexpected.

We must note, on the one hand, that the model we propose remains somewhat untested, although there is strong congruence among the variety of data sources and analyses we performed. This theory of entrepreneurial cognition presents a compelling case for explaining extremely interesting outcomes at unusual entrepreneurial firms, while providing additional validity to more established entrepreneurial theories for more mundane organizations. Equally important, this insight provides some of the only guidance for the new roles that tomorrow's entrepreneur must adopt to compete in a radically different world of opportunity and competition.

Crafting narratives

People are, simply put, not consistently rational or even efficient decision-makers. In addition, managers rarely have access to perfect information, much less the resources to perform perfect analysis on that information. Strategy and entrepreneurship scholars have spent decades arguing against traditional economic models that tend to make assumptions of perfect information and rationality. Yet these assumptions have a way of creeping back in to our research sometimes for lack of any other way of explaining entrepreneurial behavior and outcomes. In addition, the case studies we commonly use to illustrate business theory sometimes implicitly include these assumptions.

There are two reasons why we, as Western management scholars and practitioners, often default to such unlikely propositions. The first is that they are based on thousands of years of philosophical argument, codified into the very nature of important social institutions such as scientific investigation, law, and public debate. Second, the alternative to strictly rational decision-making is sometimes understood to be *irrationality*, which is clearly incorrect. There is no doubt that while intuition and emotion play potentially significant roles in entrepreneurial behavior, it would be a mistake to suggest that most entrepreneurial action is either irrational (that is, contrary to logic) or a-rational (that is, formed via non-logical modes). Entrepreneurs, on the whole, are probably no more and no less logically driven than non-entrepreneurs. It isn't unreasonable to believe that some, if not many entrepreneurs, apply different assumptions, such as the probability of success, to the decision-making process. Neither is it unreasonable to suggest that such decisions may be influenced by technology affinity, sunk costs, or other emotional factors.

So, how do we reconcile this uncertainty? Entrepreneurs are not infallible logicians; yet are not entirely emotional and irrational.

There is a complementary framework for rational logic that improves theories of entrepreneurial action. It also builds on established studies of opportunity identification and exploitation. That middle ground requires a socio-cognitive interpretation of how entrepreneurs behave. It suggests, in short, that entrepreneurs create and enact *narratives* – that is, they make sense of opportunities, resources, processes, and even their own actions, by incorporating their observations, knowledge, and even ongoing activities into plausible stories. In other words, they take the implausible, and through a combination of interpretation and action, create the plausible. Michael Lounsbury and Mary Ann Glynn argued that entrepreneurs acquire otherwise inaccessible resources precisely by functioning as storytellers.[4] Entrepreneurs present a compelling narrative to change the dominant understanding of value creation within an industry or market to access resources and shift the balance of power in their own favor.

Let's look at an example of this kind of transformation. Unlike the United States and some parts of Western Europe, most areas of India never developed wired telecommunications infrastructure. By the 1990s, the promise of cellular technology made a hard infrastructure base prohibitively expensive. At the same time, the vast majority of Indians couldn't afford monthly cellular fees charged by Western telecoms. The very companies with the capabilities to deploy infrastructure had no mechanism to monetize the cost of building cell towers across the sprawling and mountainous Indian landscape.

Enter Airtel. Founded by Sunil Bharti Mittal, this Indian company realized that the monetization would require low-cost selling

Table 3.1 *How Airtel rewrote the narrative of mobile phone service in India*

Western mobile phone industry	Airtel
Build, own, maintain cell towers	Lease cell tower bandwidth
Sell expensive phones with long-term contracts or subsidize phone costs with monthly fees	Offer inexpensive phones at cost without long-term contracts
Require upfront payments and proof of ability-to-pay	Focus on village network norms and effects to encourage compliance and payments
Expand network to meet capacity	Expand network to promote capacity
Narrative: Maximize profits and minimize risks by building infrastructure to meet demand	Narrative: Create demand for services and shift investment risk to infrastructure suppliers

techniques at the village level. Traditional electronic and direct marketing and sales were expensive and still wouldn't reach the necessary population base. Was it possible to provide mobile phone service to millions of Indian consumers without making a massive, direct investment in cell towers and infrastructure?

It was.

Airtel's implausible insight was to add a layer of logistics and administration to an already cost-constrained system. How could that work? Airtel negotiated agreements with infrastructure companies that allowed those organizations to finance tower construction and then lease capacity to Airtel. Although, in the long run, variable costs might be incrementally higher, this allowed Airtel to dramatically reduce operating costs that had to be monetized immediately. In the meantime, Airtel embarked on a village-by-village advertising and distribution campaign, providing billboards, flags, and even painting supplies to local convenience shop owners along with a low-tech turnkey system for signing customers based on limited credit and low upfront fees. Nearly every aspect of the process went against cellular phone industry standards for marketing, sales, and distribution. Table 3.1 shows how Airtel's narrative went against the dominant narrative of the industry.

Even though this initiative went counter to traditional theories of strategic decision-making, it provided a plausible narrative linking the seemingly illogical components. Airtel's insight was that even a consistent, low-cost strategic system couldn't get off the ground, because no investment plan appeared to generate the necessary returns. Instead, Airtel combined high-cost operational characteristics with low-cost but high-coverage marketing. Once the system reached scale, it would be possible to focus on creating a more consistent strategic configuration.

And it worked. Today Airtel serves more than 175 million customers – approximately the same as AT&T and Verizon *combined*. And while Airtel's revenues are dwarfed by its Western competitors, we can only imagine how that will change as the Indian economy expands and Indian income approaches Western levels. Any of the global telecoms might have implemented this implausible scheme, but they were limited by their own strategic thinking.

THE POWER OF NARRATIVE

The strength of the narrative perspective lies in the distinction between narrative rationality and logical rationality. It is very important to appreciate both the differences and complementarities between rational logic and narrative logic.

Logical rationality is formalized logic, as embodied in the logical syllogism: if A=B and B=C, then A=C. In a strictly logical framework, decisions are similarly syllogistic: any complex problem may be decomposed and analyzed step-by-step. Correct analysis at each phase ensures that the final outcome is also correct.

Narrative rationality, according to Walter Fisher, is how people actually make decisions.[5] It *differs* from strict rational logic, in that it doesn't rely on either perfect analysis or the reconstitution of deconstructed problems. But it is more accurate to see it as *complementary* to strict logic in that it incorporates a foundation of rationality within a more complicated, nuanced, and arguably realistic framework.

How does it work? What are the nuances that make narrative rationality a viable model of entrepreneurial decision-making? First, it explicitly incorporates an assessment for the reliability of the underlying data and sourcing. Can the data be confirmed? Is the source reliable? However, while these parallel the fundamental elements of the scientific method, this narrative fidelity is secondary to the critical heuristic: narrative coherence.

Narrative coherence is whether the story, as it is told, is believable or plausible. Is the explanation, in other words, a good story? Does it hang together? Does the logic provide an effective process of inference to the best explanation? Narrative coherence is, at heart, the question of whether the interlinked elements in an explanation are convincing.

Let's look at two examples to see how the power of narrative functions.

The first example is the Piltdown Man, one of the most famous scientific hoaxes in history. In 1912, Charles Dawson reported finding skull fragments in a Piltdown gravel pit at East Sussex in the United Kingdom. The reconstructed skull was widely considered to be the proof of the evolutionary "missing link" between ape and man. Skeptics of the find had emerged very early, but the forgery retained validity for 40 years. It was even referenced by Clarence Darrow in the famous Scopes trial in the United States as evidence of the theory of evolution. It was finally debunked in 1953 by scientists who confirmed that the skull was a fake constructed from a 500-year old human skull, an orang-utan jaw, and chimpanzee teeth.

Why did the hoax last so long? Part of the explanation is simply that contemporary scientists lacked a definitive age test, such as the fluorine absorption test ultimately used to expose the hoax. But a key element in the longevity of the hoax was its narrative power in the context of extant evolutionary theory. Palaeontologists had developed a theory that the human brain had expanded prior to the adaptation of the jaw; the Piltdown Man provided the critical fossil evidence to support this theory. The story was so convincing that Australopithecine fossils subsequently discovered in South Africa with conflicting characteristics were effectively ignored in scientific discourse. In other words, scientists already had a narrative in mind, and the Piltdown Man not only fitted, it helped *justify* that story. Smart, qualified individuals specifically trained in the strict rationality framework of the scientific method sometimes have difficulty applying that method when they have already adopted a plausible explanation.

The same can be seen in one of the more infamous case analyses in organizational studies: the Boston Consulting Group's (BCG) 1975 study of Honda's entry into the US motorcycle market. The study, titled "Strategic Alternatives for the British Motorcycle Industry," was presented by BCG to the British Government to explain why British motorcycle manufacturers were losing market share in the United States. The study described Honda's entry process as the result of an intentional strategy to position a low-cost, low-price motorcycle based on high-volume production and carefully targeted marketing policy.

Not quite ten years later, Richard Pascale published "The Honda Effect." He reported interviews with the actual Honda executives responsible for the introduction of the Honda Supercub and the resulting spectacular success of Japanese motorcycles in the United States. The reality was quite different than that portrayed by BCG. What had looked like brilliantly planned strategy was something else entirely. Honda's US entry process involved "miscalculation, serendipity, and organizational learning – counterpoints to the streamlined 'strategy' version."

Again, this was a case of fitting data to a narrative. BCG, as a strategy consulting firm, provides services that are primarily relevant in the context of an organization's ability to devise, plan, and implement strategy to improve performance. To the extent that the British government was seeking advice on how to help British motorcycle manufacturers regain market share and improve profitability, Honda fitted and justified the narrative of purposeful strategic action. The story of Honda's intentional, rapid, positioning-based dominance of the US small motorcycle market made sense, and was accepted despite the fact that BCG hadn't interviewed a single Honda executive involved. The story was plausible – as is often the case, narrative coherence trumps fidelity.

Two additional notes bear mentioning. First, it is no accident that Pascale referred to the cumulative picture revealed to him via the interviews as "The story that unfolded ..." Second, while one might fault BCG's method and interpretation, it should be noted that BCG's initial report was quickly assimilated into business school case studies and taught by academics as an example of effective strategic planning. The efficacy of the Supercub narrative held sway over trained social scientists as well as practitioners.

Before we apply the coherence framework to entrepreneurial business models, let's clarify how narrative coherence works. Once that framework is in place, we can debunk the false promise of complexity, and consider simple examples of narrative coherence. We'll then look specifically at examples of how entrepreneurs utilize coherence, both to maintain stability and effect organizational change.

COHERENCE, NOT LOGIC

For the purpose of consistency, we will focus primarily on Fisher's definitions and application of coherence. Fisher distinguishes between the "rational world" paradigm, which fits relatively well with neoclassical economics and theories of strategic planning, and the "narrative world" paradigm, which forms the basis for our insights into entrepreneurship.

In the rational world, people are predominantly or entirely logical. Science, accumulated knowledge, and circumstance provide the basis for argumentation. Decision-making is the result of

Table 3.2 *Rational vs. narrative perspectives*

Rational perspective	Narrative perspective
People are rational	People are storytellers
Decision-making based on argument and logic	Decision-making based on "good reasons"
Argument driven by science, law, circumstance	"Good reasons" driven by experience and language
Rationality comes from knowledge and rhetoric	Coherence is constantly tested against observation
The world is a set of logical puzzles to be solved	The world is understood as a set of stories

Adapted from Fisher, W. R. 1985. The narrative paradigm: an elaboration. *Communication Monographs*, 52: 347–367.

that argumentation, solving the puzzles that comprise the human experience.

In the narrative world, people are essentially storytellers. They apply their experience and character to develop plausible explanations or "good reasons" that inform decisions. Those decisions are then evaluated based on how well they fit the ever-evolving stories that make up our lives. It is important to note that by "storytellers" we don't refer solely to the communication of narratives from one individual to another. In the narrative framework, storytelling refers to the underlying sense-making process we use to understand the world. In other words, the act of living involves an ongoing process of telling and interpreting our own stories, both to ourselves and to others. This has special connotations for an organizational context. A firm may thus be viewed as a multi-vocal community in which individual, group, and organization-wide stories are being interpreted and told. Table 3.2 explicitly contrasts the rational and narrative perspectives.

NARRATIVE COHERENCE: PLAUSIBILITY AMID UNCERTAINTY

What then, is coherence, or plausibility? It is the preponderance of evidence – the explanation that best fits the key underlying information without significant conflicts or logical flaws. Here are the primary characteristics associated with the development of a coherent narrative:

- Minimizing, rather than eliminating, inconsistencies
- alignment with the preponderance of evidence, rather than the perfect explanation of all data
- local coherence, or grouping of consistent elements to isolate inconsistencies.

At the surface level, we can see how an organization utilizes coherence during adaptation by looking at Return Path. At Return Path, the world's leading email whitelist company, coherence stems from the firm's core capabilities in email deliverability and the organizational emphasis on human resource development. Return Path has been through at least three distinct business models in its 12-year life in the email deliverability space. The initial business model associated with email forwarding services expanded into email marketing campaign services and then ultimately evolved into whitelist-based deliverability services.

During these organizational change processes, some of which involved acquiring or divesting technologies or entire businesses, CEO Matt Blumberg relied heavily on the firm's core focus on people-centric work and the vision of "rewarding good email senders." Across the organization, regardless of business function, seniority, or tenure, employees openly discuss being "the good guys" in the war on email spam. And so the organization retained coherence despite dramatic shifts in personnel, product focus, and customer interactions. As one example, an early incarnation of Return Path acquired a technology from another company run by George Bilbrey, who stayed briefly during integration but then left to work on other projects. Amazingly,

Return Path then acquired George's next venture after helping to partially incubate it, bringing George back on a permanent basis. The people, processes, resources, and circumstances underwent significant modifications, but the employees present during these transitions commented that the company never changed.

Coherence supersedes small changes, even when the sum of those small changes theoretically adds up to change in the whole. Readers of fantasy might recognize this from Terry Pratchett's explanation of dwarvish axes handed down through the generations. Once in a while the handle may rot and need replacing, or the blade loses its edge and is swapped out, but it's still the same *axe*. This is known in philosophical studies as the Ship of Theseus paradox, described iconically by Plutarch:

> The ship wherein Theseus and the youth of Athens returned [from Crete] had thirty oars, and was preserved by the Athenians down even to the time of Demetrius Phalereus, for they took away the old planks as they decayed, putting in new and stronger timber in their place, insomuch that this ship became a standing example among the philosophers, for the logical question of things that grow; one side holding that the ship remained the same, and the other contending that it was not the same.
>
> (*Theseus*, Plutarch – The Internet Classics Archive)[6]

The paradox appears much less problematic for the organization compared to the ship. A variety of formal and informal structures may remain constant at the firm, including legal structure, facilities, formal hierarchy, and even mission, culture, and standard operating procedures, even as all the resources, activities, and ideas change. But coherence at Return Path incorporates more elements than just the legal boundaries of the firm. The retention of core principles, down to Blumberg's tradition of meeting with individual employees at all levels in the organization for informal conversations and open roundtable forums regularly hosted by all executives, facilitated cohesiveness and promoted plausible, common narratives more effectively than a set of incorporation documents.

In addition to minimizing inconsistency and employing local coherence effects, strong narratives attain plausibility amid uncertainty via partial overlook of flaws and establishing goal accessibility. These tools are utilized by entrepreneurs building businesses that may achieve the unexpected.

Coherent narratives enable actors and organizations to partially overlook problems. In traditional frameworks of organizational strategy, inconsistencies are forms of organizational weakness. Firms with complete systems of mutually reinforcing internal elements generate the greatest competitive advantage. But in the entrepreneurial context, especially during novel innovation processes, that advantage is not as clear. Some of the most extraordinarily innovative firms simply don't know the ultimate market context in which they will compete, or even how the actual product or service will be embodied. Even when the technology and market dynamics appear well-established, many entrepreneurial firms benefit from uncertainty in organizational structures in the process of enacting opportunities.

There are two reasons for this. The first is that low levels of conflict help retain options when outcomes are unpredictable, Second, coherence presents a thought process so familiar that we apply it without even thinking about it.

Consider the case of Metalysis, a Cambridge University spin-out commercializing a novel metals processing technology. In the early stages of commercial development, the company faced a potentially significant challenge. A variety of high-value metals are currently refined via the Kroll process, a 50-year-old technology that is capital and energy intensive. Although the Kroll process is not inherently ecologically problematic, the transport, extensive chemical requirements, and processing footprint associated with refining millions of tons of raw ore present unavoidable environmental hazards. Metalysis' FFC Process™ offers dramatic improvements through a low-temperature energy cell that works a bit like a battery in reverse. Electricity flows through the cell containing the ore and specific reagents; the target metal molecules are effectively separated from the ore impurities.

Metalysis had made excellent progress on perfecting the process cells for titanium (Ti) and tantalum (Ta). The conundrum lay in the question of scale. Tantalum production is generally estimated at less than 2000 tons per year. Titanium production, on the other hand, is 60–80 times as great. To be a viable participant in the titanium industry requires significant scale capabilities. The properties of the two ores and the refined products present significant physio-chemical differences. Titanium is primarily targeted towards high-strength structural applications such as aerospace, while tantalum is predominantly used for small quantity, high-value applications such as electronics components in mobile phones. On the whole, titanium producers are not also tantalum producers, because the processes operate differently and on dramatically different scales.

What was Metalysis' business model narrative? Would it license its technology to producers? Build captive refineries? Build pilot scale refineries and then contract out the scale-up process? Partner with established refiners? This challenge goes beyond investor pitching. The question was on the minds of employees, because the decision to prioritize one metal over the other would have significant implications for human resource management across the organization.

The narrative at Metalysis was initially focused on the innovative technology, similar to many or most early stage technology firms. But that narrative has reformed around cleantech innovation and long-term scale-up possibilities without actually converging to a specific market segmentation process. This is, by most strategic and entrepreneurial assessment frameworks, problematic or even dangerous to the survival of the firm. We actively teach lean development and resource commitment to entrepreneurship students, precisely because indecision or "spreading" of goals seems to diffuse the firm's focus.

But Metalysis has found a coherent narrative in the gray area between technology development and market entry, precisely because global markets and industry dynamics in high-value metals fluctuate dramatically year to year. The wrong bet could be costly for a large organization, but fatal for a small one. In addition, the developmental narrative in place at Metalysis allows it to overlook minor inconsistencies in the business model. For example, the organization has been able to shift resources and processes across research and application areas at relatively low cost. The inherent unpredictability of scientific research and testing don't always permit lockstep planning and implementation. These rearrangements and realignments, if mapped onto an organizational chart or process diagram, would show up in traditional assessments as wasted efforts or duplication, because more explicit work processes driven by well-identified market goals would theoretically provide shortest-path procedures.

Instead, Metalysis management retains the flexibility to adjust priorities and resource utilization week to week. As one of the product managers explained, "We can shift experiments within the labs as needed. Today those cells are working on a project, but when the run is over, we may have an entirely different process test to implement. The goals at the group and product level allow us to make those changes on the fly." Without doubt, as the company enters markets at scale, a new narrative will be needed – perhaps one that solidifies more clear business model structures. In the meantime, however, the Metalysis narrative has raised more than £25 million in investment funds from the most sophisticated venture investors, including ETF in London.

The partial overlook of flaws is linked to another critical characteristic of coherent narratives: goal identification and accessibility. Insight 2 helped lead us to goal maps, which show how firms like Confederate Motors can adjust activities and resources quickly and smoothly without contradicting goal prioritization. The plausibility of a narrative, however, depends in part on whether the goals are understood to be *accessible.*

By accessible, we don't mean that they have to be near-term achievable or reasonable. Accessible means that the actors, whether employees, managers, owners, funders, or customers, can understand and embrace those goals as valid and appropriate. Is Confederate's goal of invigorating American industrial design reasonable or near-term achievable? Objectively, this could only be assessed in a long-term historical context. But to appreciate the power of goal accessibility we have to step out of the strict rational framework and into the narrative rationality at Confederate.

First, the company presents a plausible argument for Confederate's impact, and that argument is aligned with employee motivation and the customer-facing elements of the firm. Arguably, Confederate does, in fact, participate in the conversation on industrial design. Confederate motorcycles have been lauded and panned in a variety of design-oriented magazines more commonly dedicated to clothes, architecture, and art. And Confederate has influenced broader conversations within the heavy manufacturing industry. The Confederate Fighter was the only motorcycle at the 2010 New York Auto Show. And the radical oversized front forks on the Confederate Wraith have now been imitated in a variety of concept models at other manufacturers. But within the organization, these examples are not considered supporting evidence of the firm's effect on design; they are consistently reported as *results* of Confederate's influence. The narrative effectively becomes the null hypothesis.

Second, Confederate's industrial design is tightly linked to the element of rebellion. The very fact that the firm's radically retro emphasis will not, in fact, be globally embraced, plays perfectly into the firm's David vs. Goliath motif. Confederate's narrative is precisely that of the perpetual underdog engaged in an impossible battle, winning by unexpected means. In other words, the *lack* of success is also interpreted as confirmation of the narrative. And this story, especially tied into the biblical narrative, is instantly accessible to stakeholders.

There is a fine line between building an effective narrative with both supporting and conflicting data, and simply choosing to live in cognitive dissonance. Our research has not clarified the specific characteristics that suggest when the crossover happens, but there do seem to be some clues. Comparing Confederate to Savage Entertainment helps suggest what those clues are.

First, human familiarity with narrative structures means that all of us are highly sensitive to even small inconsistencies, or plot elements that point towards negative rather than positive reinforcing cycles. At Confederate, employees at all levels were emphatic that the firm's design-led focus and uncompromising attitude were critical elements in the success of the Wraith, regardless of the final sales figures. At Savage, on the other hand, there was a clear perception that the company's battle was noble but ultimately unwinnable. Contrarians might argue the Savage executives could have instilled a different interpretation. Perhaps this is possible, but our instinct, in fact, is that something quite different is happening. Charisma, as displayed by Matt Chambers, Raj Dutt, or Pramod Chaudhari, is clearly a valuable entrepreneurial talent. And it plays an important role in narrative formation and evolution. But the Savage employees were responding, with hesitance and trepidation, to the impending collapse in the broader narrative of the organization as a whole. They saw the inconsistency, more and more clearly, and they could link it to a longer-term narrative at the organization: fighting the good fight.

Second, identifying different narratives within the organization is not necessarily indicative of fundamental dissonance, but difference in the patterns of narrative is a good litmus test for dissonance. At Confederate, not every employee embraces Chambers' objectivist humanistic philosophy and the implications for the organization. But differences vary based entirely on individual personalities. At Savage, differences in narrative coherence were pronounced primarily by seniority. High-tenure, high-seniority employees presented the most dissonant perceptions of coherence. Low-tenure, low-seniority employees registered uncertainty but no significant dissonance. This example is indicative, not definitive. We have heard anecdotal stories about firms where the newest and most junior employees were most likely to see the inconsistencies in the business model. In the case of Savage, it appears that seniority and tenure provided the necessary high-level perspective on industry dynamics, which was the crucial indicator of whether Savage was ever going to develop novel creative properties.

It does appear, in rare cases, that narrative collapse can be averted, or organizational dissonance tolerated via combinations of charisma and questionable activities, as may have been the case at Enron and Sunbeam, or the investment community around Bernie Madoff. The role of pure dishonesty, or fraudulently based opportunism, undoubtedly plays a role in narrative coherence at organizations. We did not observe this type of behavior, and have no reason to suspect such was the case at any of the organizations we studied. Without exception, entrepreneurs were pursuing the unexpected with the highest levels of personal and professional integrity. So, for now, we can only speculate on the specific effects and interpretation of fraudulent behavior on coherent business models. Our instinct is that the high sensitivity of human perception for narrative inconsistencies suggests that individuals are likely to perceive the dissonance, but that narrative can be turned to less salubrious purpose by enabling rationalization and objectification. This fits with other research on dishonesty in local workplace contexts,[7] but this is as far as our own analysis may proceed.

Narrative coherence thus provides an effective explanation for how entrepreneurs develop stable organizational systems in the pursuit of implausible opportunities. What is even more interesting, however, is how entrepreneurs manage narratives of change.

THE NARRATIVE OF CHANGE AT INNOVATIVE ENTREPRENEURIAL COMPANIES

Relatively little research has considered a narrative framework for understanding organizational change processes, despite the massive literature on behavioral psychology and sociology describing and addressing decision-making and behavioral outcomes. Entrepreneurship research has demonstrated that dynamic capabilities for reshaping routines and reconfiguring resources are especially important in dynamic and uncertain environments. For example, early stage firms are more likely to use trial and error and improvization, generating dramatic change based on learning from action rather than learning from planning.[8] But managing narrative as a key dynamic capability that directly influences individual behavior, firm-level learning, and structural configurations hasn't received significant attention.

Our research at innovative entrepreneurial companies specifically considers how entrepreneurs manage narratives of change, especially radical change associated with business model innovation. Let's take a look at some of these change processes within a framework of narrative coherence at our study companies. Our observations suggest that there are at least four types of narrative transition associated with successful business model change at entrepreneurial firms: growth, extension, focus, and displacement.

Narrative growth at SustainableSpaces

A lot of attention in both the popular press and the research community has centered on mission-driven businesses. The nomenclature of double, or triple bottom lines has given way to "social entrepreneurship." This appears to refer to entrepreneurial activity that places a premium on outcomes related to social goods, such as welfare and ecology, rather than strict profit-making. In fact, as the goal mapping exercise associated with Insight 2 demonstrated, all firms place priorities on a variety of outcomes. But it would be a mistake to suggest that firms develop narratives associated with multi-level goals and then never change either the goals or the narrative. There is little doubt that many entrepreneurial firms are initiated within a primarily profit-seeking mode and then adopt, evolve, or co-opt alternate outcome priorities. Similarly, many organizations with explicit orientations away from profits find, willingly or not, that organizational sustainability often requires an emphasis on cash, and even profit-management.

Innovative entrepreneurial companies present fascinating case studies of complex and interlocking goals and processes. One of our study companies, Recurve, clearly fits this mold, and offers the opportunity to understand not just the narrative of social entrepreneurship, but the growth of a more complex narrative at a mission-driven company.

Recurve's founder, Matt Golden, makes it absolutely clear his objective has always been to impact energy consumption and petroleum dependence associated with residential energy consumption. From selling solar panel solutions, Golden became convinced that a one-size-fits-all perspective to residential energy efficiency was suboptimal. Government incentives were transient and confusing, encouraging buying behavior that would, in the long run, perform more poorly than other energy management modalities. He formed SustainableSpaces to provide experienced, specialty audits of home energy efficiency to address this, on the assumption that it was better to solve the problem correctly one house at a time than incorrectly for the mass market.

Ironically, this enactment was suboptimal for Golden precisely because it worked. It couldn't be scaled because all the expertise was held tacitly within the organization and more specifically in his experiential knowledge. That tacit knowledge helped Golden and his auditing engineers identify best options for residential energy improvements for each property, but those solutions couldn't be implemented beyond each idiosyncratic audit. Golden also knew that SustainableSpaces' success was in part a function of the unique characteristics of the local region. The San Francisco Bay Area combined higher-than-average consumer interest in ecological issues, higher-than-average property values, and residential construction norms that did not emphasize energy efficiency.

For Golden, running a bespoke residential energy efficiency auditing service was necessary but not sufficient. Impact had to be measured at a broader scale. A different business model would be

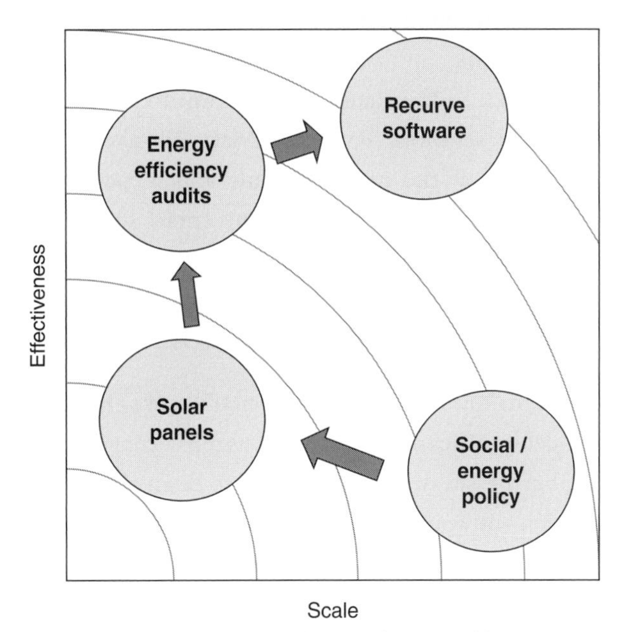

FIGURE 3.1 The evolution of Matt Golden's energy impact narrative

required, one explicitly designed to be scalable, even if it meant the application of dramatically different capabilities.

The evolution of Golden's narrative of impact is shown in Figure 3.1. The changes modulate the narrative between "effectiveness" and "scale." In this narrative, the organization is, in fact, just one of many heterogeneous elements that comprise a configuration of coherence. Information is obtained experientially, and new elements are brought in and new processes initiated in the search for a coherent solution that meets functional and mission objectives. For Golden, the story of the organization is a narrative of accumulation and growth in which the configuration becomes more complex at each stage passing through suboptimal stages in the process.

Narrative extension and narrative focus at Return Path

Return Path, discussed in detail in Chapter 1, presents a very different profile. Although the narrative at Return Path has changed throughout the firm's evolution, evolution has alternated between periods of extension and focus.

As a start, we go back to the firm's early efforts to sell email change-of-address (ECOA) services. Founders Matt Blumberg and George Bilbrey had guessed at the ultimate scale and scope of the email market, and the resulting need for updating email addresses. But the story wasn't holding together in the context of monetization. The solution required extending the business model narrative, exploring links between multiple markets, as Blumberg describes:

> We still wanted to make a run at the ECOA market. We changed the business model significantly from being purely a cooperative network among businesses, which had proved to be an all-or-nothing proposition – it wouldn't work unless we had a thousand companies. Instead, we decided to view the business as having two distinct sides to it: a consumer side and a corporate side. We treated them each differently. We went out to get the data from consumers directly, and then we tried to build a sales channel into companies, leveraging one market as a way of getting into the other. It started working – we went from no revenue to some revenue.

Ultimately, however, even this extension wasn't sufficient. A few million in revenue was not enough, especially for a firm that had taken over $10 million in venture capital at that stage in the game. The narrative of "some success" wasn't fully coherent with the perceived opportunity and the goals of the stakeholders. The problem now, as Blumberg described, was that the original business model core, ECOA, was holding back the potential of the broader email deliverability platform:

> We reached that fork in the road, the board and management team had a series of heart-to-heart conversations revolving around the choice of continuing to focus on ECOA – not so interesting – or just calling it a day, selling the ECOA business

> and losing a little money on it but hold our heads high that we accomplished something, or viewing the company, its employee and knowledge base and clients as a platform to do other interesting things around email and view this as a "restart."

The extension of the narrative had, on the one hand, provided a means of creating some success, but at the same time clarified that "some success" simply wasn't a preferred outcome. Again, the firm's process utilized an extension, this time a more dramatic one. It's probably not surprising to see firms utilize the same narrative change processes repeatedly – like all processes, these may become familiar and routinized. Blumberg explains:

> We knew our core assets were expertise around email, a client base of blue-chip companies like eBay, AmEx, Sprint, and Microsoft, a great team, good investors, and deep pockets. Our theory became: let's offer a lot of things that have a common thread of email marketing – let's offer everything other than email delivery, which is already a hugely competitive sector. We were going to have an enormous amount of customer leverage. We knew up front we wouldn't have much technology or operating leverage, because we knew they were different things, but we knew customers would buy multiple products from us. We decided to become the company that does everything email other than deliver, so we entered into a series of acquisitions and some organic growth.

As might be expected, this diffused the efforts of the organization – the business model lost coherence as the small company attempted to serve multiple products to multiple markets. Even so, the power of narrative can be seen in the mindset of Blumberg and the other key executives. Blumberg commented, perhaps somewhat wryly, that at one Board meeting someone referred to the organization as "the world's smallest conglomerate." This demonstrates organizational sense-making in real-time and shows how fully integrated the narrative framework is within the human experience. Narrative formation is not fully isolated in individuals, because interpretation almost always takes into account the perception of others' understanding. Despite this, alternate narratives can explain the same data. In this case, the reality check came from another director, as Blumberg recalls:

> The businesses were so different, it was mind-bending for executives. It was like we each had three different jobs. Finally, one Board member said: "You guys are running two start-ups and a turnaround."

It's important to understand that the company was, in fact, succeeding – with revenues accelerating from about $2 million to almost $20 million. But it was working despite the fundamental conflicts inside the organization and the realities of the market. The company was holding together, but the coherence was slipping. Blumberg summed it up: "The business model just didn't work. There was no customer leverage via cross-sell, at all. Although many customers bought multiple services from us, they had completely different buyers, intermediaries, and decision cycles for each service."

The Return Path team made the conscious decision to dismantle the complex system that they'd put in place over nearly five years. The company split off extraneous businesses after unsuccessfully trying to sell some of them. The narrative had to be focused: the email world was bigger than ever, but even a firm as aggressive as Return Path couldn't serve all of it. The executive team and Board felt they understood the core piece – the link between prior sending success data and future deliverability. It's the same scale effect that Google has been leveraging in developing and launching services like automated translation of websites from one language to another, based on prior translation data. The team began referring to the core deliverability data set as "the machine of infinite possibilities." The firm already had access to email receipt data from hundreds of millions of emailboxes via Hotmail and other organizations, and the decision was made to drive investment in that asset.

The focus in narrative can be seen explicitly in Blumberg's language, as he describes the evolution of the model:

> This was a fundamental change in architecture of [the] business model. But it follows material change in the architecture of the email industry. It was led by opportunity, partly by perceived future opportunity. The spam problem is what has us in business in the first place. Ten years ago, spam was 25% of email. Most spam filtering companies started at that time. Spam is now 95% or more of all email traffic. The spammers are getting sophisticated. 60% of spam is coming from individual computers hijacked by viruses, and so spam is getting harder to spot. We have to assume spam will approach 99.99% of email, so the more interesting problem is not finding what's bad, since almost all of it is bad, it's identifying what's good. We're the only company focused on identifying what's good. Finding the bad email is basically participating in an unsolvable nuclear arms race with the spammers, but finding the good email is a very tricky, but solvable problem.

There are a number of narrative elements to notice here. First, Blumberg has dramatically narrowed the model to a specific problem – identifying non-spam emails. Second, there is a more clear rationalization of the link between the firm's target opportunity and change in the industry – in other words, the model changes to match exogenous change. Finally, the heroic narrative has begun to emerge – the organization is on a quest with uniquely positive qualifications. The coherence of this type of story is especially powerful, as it incorporates archetypal elements of narrative that are entrenched in human culture.

One of the wonderful characteristics of narratives, and especially narrative models of opportunity, is that they are dynamic. Even as Blumberg and his team at Return Path have focused the various organizational narratives down to an extremely compelling story, changes have begun to emerge:

> We have to provide software plus service – whitelisting and reputation management. But the really interesting developments will be solving the many-to-many problem, moving down the long tail. Right now we work primarily with the large email senders like eBay, helping them get through the big email aggregators like Hotmail, but ultimately this could be the equivalent of [Google] AdSense. The trick now is to serve the relatively small number of big guys with huge email traffic as well as the monstrously large number of firms with relatively small traffic.

Blumberg wants the company to change email management for small players. It would be an embedded mechanism to help any email sender, big or small, ensure that valid emails get through, whether sending a dozen emails a day or a million every hour. It all seems logical now, perhaps, a token to the power of narrative and rationalization. But it is that same power that has propelled the company through extraordinary change.

Narrative displacement at Savage

Perhaps the most interesting narrative shifts in entrepreneurial business models are displacements. These can develop for a variety of reasons: distress, surfeit, opportunity, and even confusion. Displacement is the uptake of business model narrative elements that intentionally or unintentionally replace other elements. Displacement is quite different from growth, extension or focus. Those processes function in a continuous and organic mode, in which the new equilibrium can be linked to the old by coherent connections. Growth and extensions build on a retained base of elements, while focus processes excise apparently extraneous elements. Displacement, on the other hand, discards old elements in favor of new elements. Although threshold effects are potentially relevant for all types of narrative change, displacement shifts are more likely to result in incoherence and other discontinuous outcomes.

At Savage Entertainment, the video gaming company, the founders always planned to develop original content funded by contracted project work. One of the most formative events at the company, however, dramatically changed the operational model. For the first three years of the company's existence, Savage worked on one project at a time. There were a variety of narrative rationales associated with this. The founders continued to research and develop ideas and enabling technologies associated with original IP development. The firm promoted a close-knit culture that actively retold the plot of the founders leaving Activision to develop original product. The company was one team, working on one project at a time. And, since new product could, at the time, be developed with about 10–20 people, there was no clear value to expanding the organization beyond that scale.

This narrative became incoherent in the wake of crisis: the customer that had contracted for Savage's largest development project reneged on payment. It took nearly a year to pursue the payments via legal recourse, during which time the founders repeatedly considered shuttering the company. In the end, payment was recovered, and the founders displaced a key narrative element at the organization – the single-project node. As founder and CFO Chacko Sonny explained:

> We never wanted to be in that position again – entirely dependent on one customer – it just seemed too risky now. We would have never guessed that a customer would just arbitrarily refuse to pay for services already provided – for product already delivered. But it happened, and we changed our minds. From then on, we always ensured we had multiple projects in production, and multiple projects in the pipeline. Losing a project or payments for a project would hurt, but not kill us. We put survival ahead of everything else.

It may not have been obvious at the time, but while this displacement was logically consistent with the firm's other practices, it wasn't fully coherent. In theory, there was no reason that a multi-project firm couldn't still develop original content. In fact, expanding the project portfolio resulted in more capabilities imported into and developed at the business, and the added stability could be beneficial to assigning slack resources to more creative works. But the narrative shift was dominant. Before the crisis, the emphasis on sustainability and project outcomes were outweighed by the intent to develop original game content. But the new focus on ensuring survival carried a narrative cost – it required that the performative aspect of the organization's business model de-emphasize work associated with the organization's long-term creative interests.

The problem, perhaps, was that the founders did not change the explicit business model. They continued to tell the organizational story as if the firm's primary goal was the development of original, creative content. This story was so entrenched at the organization, and retold continuously over time, that it was echoed in interviews seven years after the crisis by employees only recently hired:

> Obviously we have to have contract projects to pay the rent, but we're always looking towards developing our own titles. Chacko and Tim have been working on that – it's one of the reasons I joined the company, to have the opportunity to do creative work on new games eventually.

But not every employee believed this. Some of the longer-tenure employees clearly had accepted the performative reality and assimilated that into their interpretation of the narrative business model:

> Sure, we keep talking about doing our own games, but unless something unexpected happens, I don't think that's going to happen. We're a job shop, really – we do it faster and cheaper on short notice. We're good at it, but we don't do the really original stuff.

How long could this narrative dissonance be maintained? Savage's successful 12-year run ended when changes in industry capacity management and the economic downturn combined to create a trough of business deeper than Savage's project portfolio model could ride out. But the strain in the organization was already apparent. After all, Savage was, theoretically, a game developer that had never completed development of a novel, self-contained game. As one employee noted, just six months before the industry felt the full brunt of the economic crisis:

> It's kind of an endless losing battle, in a way. Chacko and Tim are constantly trying to get money from customers, drum up new business and projects, but we never seem to work on the really interesting stuff – it's all short term and urgent.

Displacement doesn't have to be a negative business model change. But because it is a discontinuous change, the rationale for change and the effects on organizational coherence are likely to be unpredictable.

MODELING NARRATIVE COHERENCE AT CDI

We've documented four types of narratives for business model change – growth, extension, focus, and displacement. When companies drive towards achieving the unexpected, narrative change needs to result in coherent structures aligned with the opportunity and design for exploitation. Applying the same modeling structure we developed in the discussion for Insight 2, we're going to provide a detailed analysis of remodeling organization design for coherence.

Cellular Dynamics (CDI), based in Madison, Wisconsin in the United States is a world leader in stem cell technology research and development. The company was founded in 2004 by a group of scientists, financiers, and professional managers. The team included Dr. James Thomson, the University of Wisconsin-Madison scientist credited with first isolating primate and human stem cells. The other key founders were Dr. Thomas Palay and Mr. Robert Palay, the managers of Tactics II, LLP, a venture fund specifically formed to invest in the commercialization of stem cell-related technologies. CDI licensed relevant human embryonic stem cell technology developed in Dr. Thomson's laboratory from the Wisconsin Alumni Research Foundation (WARF), the technology transfer entity associated with UW-Madison. Management anticipated that CDI would develop assays based on stem cell technology to support or accelerate development of pharmaceutical therapeutics.

A second entity, Stem Cell Products Incorporated (SCP) was created in 2005. License rights were carved out from the original WARF licenses to enable long-term research to develop novel therapeutic compounds. The initial therapeutic target was the development of stem cell-based red blood cells or platelets that could be produced in large volumes for transfusion or other related applications.

A third entity, iPS Cells (IPS) was created in 2006 to serve three purposes. First, the firm licensed a new and potentially disruptive technology from the same university. Second, the company would assess out-licensing opportunities for that technology and other technologies owned by CDI and SCP. Finally, the firm would investigate in-licensing of other technologies related to scale-up and commercialization.

Separate entities had been formed for a number of reasons. The founders believed that the firms would utilize distinct revenue models associated with different product characteristics. In addition, each entity had a different founder set, creating non-obvious equity valuation issues. According to President Tom Palay:

> [A]t the time of founding, the founders perceived that the businesses had different business models – so there was a concern that the high-risk, high-reward company could damage the lower-risk company ... It was mathematically easier to have separate entities, especially since we didn't know what was going to happen to the separate businesses.

At the same time, Dr. Thomson and the Palays were central to all three organizations. The entities also shared key operational

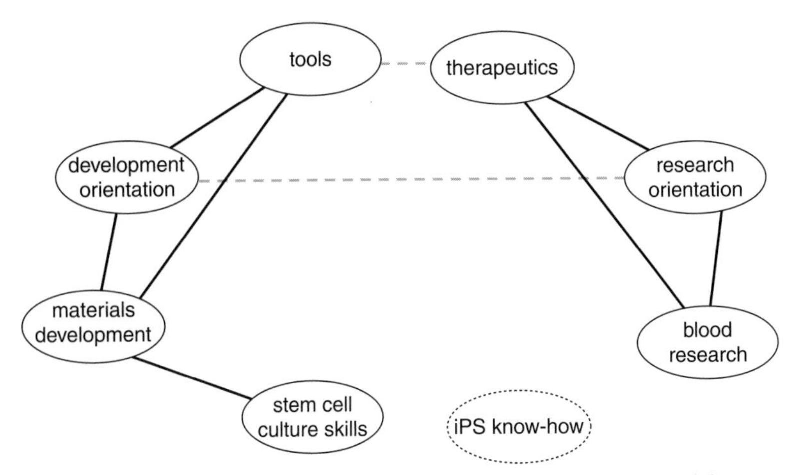

FIGURE 3.2 Elements and interactions in CDI's business model

managers as well as central administration and some physical facilities. As of 2007, the point of the firm's "initial configuration," the combined entities had raised more than $15 million in venture finance, and had a combined headcount of approximately 20 full-time employees.

A fundamental lesson on achieving the unexpected is that applying typologies or generalized structures to business model analysis adheres to traditional strategic management thinking. Applying previously identified organizational patterns to entirely novel situations may ignore the power of narrative rationality. Therefore, to simulate the narrative business model, we used the specific information from the extensive interviews across the organization. We assessed the language and words managers and employees used to describe the business model, both explicitly and implicitly. Based on resource, transactive, and value structures, we developed a simple, but multilayer network describing the most critical elements in the business model of CDI. That model is shown in Figure 3.2.

The dark lines represent mutually enhancing relationships, while the light lines show conflicting or inhibiting relationships

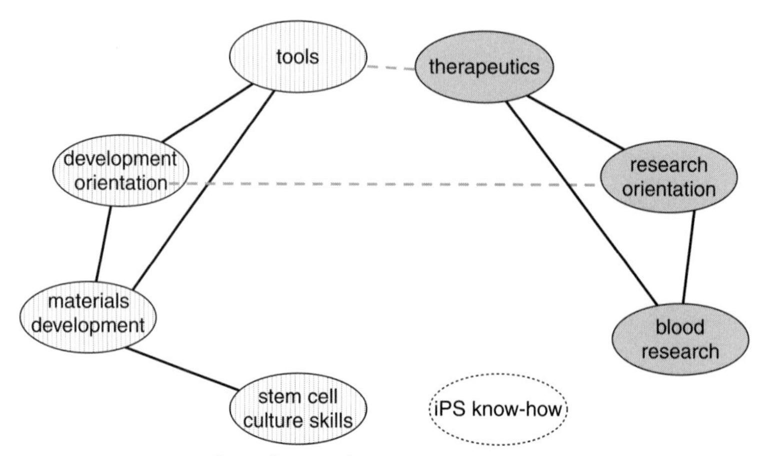

FIGURE 3.3 Less than perfect system at CDI

between elements. The thickness of the lines represents the relative strength of the relationship. This network can be simulated using a neural network. The elements exist in binary states, and "running" the network occurs as the simulator updates the state of the elements. For example, the simulator selects one element at random, examines its state and relationships with neighbors, and decides whether changing the state of that element better satisfies those relationships. Over thousands of updates, configurations of elements emerge. If a single configuration emerges from repeated runs of the simulator, we say that configuration is the stable solution.

In the case of CDI in the initial configuration, the stable solution is shown in Figure 3.3. The simulator has generated the multi-structure configuration that matches the division between the tools and therapeutic businesses. While this solution is stable, it still incorporates numerous conflicting relationships. These are, realistically, inevitable in most organizations.

At the time of this initial configuration, the separation of the entities was acknowledged as "imperfect." It was justified with both positivist and negative narratives. The positivist narratives included the complexity of satisfying equity interests of distinct founder groupings and distinguishing between business models for potential funders. In contrast, the negativist narratives diminished the perceived problems by noting that shared management teams and facilities were cost-effective and encouraged communication between groups. David Sneider, the chief business officer at the time, commented that "the only explicit cost is associated with accounting for resource use, and we have computers to make those calculations for us."

Disruptive technology acquisition

In 2006, Thomson's research at the University of Wisconsin-Madison proved the potential for induced pluripotent stem cells (iPS). The third entity, iPS Cells, Inc., was formed specifically to license this technology from WARF. This was a potentially important development, because iPS methods generate stem cells from adult rather than embryonic cells. This advance resolved a number of outstanding technical and operational issues associated with stem cell production and commercialization. In addition, the use of cells derived from embryonic sources had been a continuing source of ideological controversy.

The uptake of iPS technology impacted numerous functions at CDI and Stem Cell Products. It changed the conceptual and narrative frameworks utilized by executive management as part of operational and long-term planning. The combination of the new technology and the new narrative altered the underlying capabilities set at the organization as well as management's interpretation of the business model for commercialization. The initial configuration focused on assay development for drug discovery. By 2009, the narrative of the firm's core resource base had apparently *regressed* to a focus on cell manipulation. As Tom Palay noted in 2009, "Our expertise is in the automation and production of cells based on culturing stem cells and differentiating those cells."

This adaptation completely changed the organization's long-term manufacturing strategy and capability requirements. At the time iPS was licensed, the company needed to resolve a variety of technical challenges to large-scale, high-efficiency stem cell manufacturing. Variations in cell culture stocks from vendors and extant culturing processes limited scale-up of fully differentiated stem cells. iPS provided the possibility to move to a unified cell-type manufacturing platform based on a cell culturing framework leveraging a more limited cell stock. In addition, the platform provided a common basis for stem cell competencies associated with drug discovery tool products and long-term therapeutic product development. At a tangible resource level, the iPS platform simplified materials development and provided a consistent basis for the development of necessary but nascent automation skills.

In Spring 2008, executive management and the founders of the separate entities held a series of strategic meetings. The discussions concluded that iPS provided a common platform for longer-term technology development for both therapeutics and tools. Prior discussions about merging the organizations had focused on ensuring "fairness" with regard to equity stakes, as each entity had a different set of founders. Stock transfer pricing was a potentially volatile topic difficult to resolve to the satisfaction of all parties.

Two other financial factors appeared to influence the transition process. First, a large federal grant application to fund long-term therapeutic research was unsuccessful. Second, executive management was unable to secure venture financing from a lead venture capital fund. Closing a venture round could not be anticipated in less than six months from initiation of new investor discussions, and the market for venture financing had significantly contracted.

In numerous interviews, managers expressed surprise and resignation about financing. Some were concerned that the complex structure incorporating apparently conflicting business models confused financiers. By mid-2008, management had acknowledged that a significant, near-term venture capital investment was unlikely.

Traditional strategic management points towards separating business models and value chains to ensure alignment and consistency among organizational elements. Although the administrative costs of shared systems were relatively low, the conflict in market-facing business model structures appeared to limit financing and diffuse operational focus. Separating the entities would be in line with theories of core competency, transaction costs economics, competitive strategy and advantage, strategic complementarity, contingency and configurational strategy, and risk portfolio analysis. The simplest tradeoff appeared to be the relatively minor cost of contracting among friendly parties against the apparently significant cost of conflicting internal business systems.

But, in Summer 2008, executive management recommended *merging the entities*. The merger was announced in late 2008 and implementation lasted into the summer of 2009. A variety of structural and cultural changes merits further discussion. The research team from SCP was significantly redeployed to development work. Of 20 scientific researchers only five were retained to continue long-term research on therapeutic products. The general manager of iPS, Inc. had been serving as chief technology officer of CDI and SCP. Following the merger he retained direct oversight of the five long-term researchers, but was redirected to greater focus on acquisition of enabling technologies. Direct oversight of long-term R&D was shifted to a research manager within the merged organization.

Figure 3.4 shows how the addition of the new technology changed the network of relationships in the business model. This includes the interactions between the new technology element (iPS know-how) and the prior elements.

Figure 3.5 shows the resulting stable configuration based on the simulated search for coherence. The color of the elements reflects a stable configuration that emerges from this updated system. Unlike the prior system which clearly demonstrated two distinct subgroups, this system reflects a single large group that incorporates all but one of the elements. It incorporates the conflicting interactions of research and development orientations as well as that of tools and therapeutics.

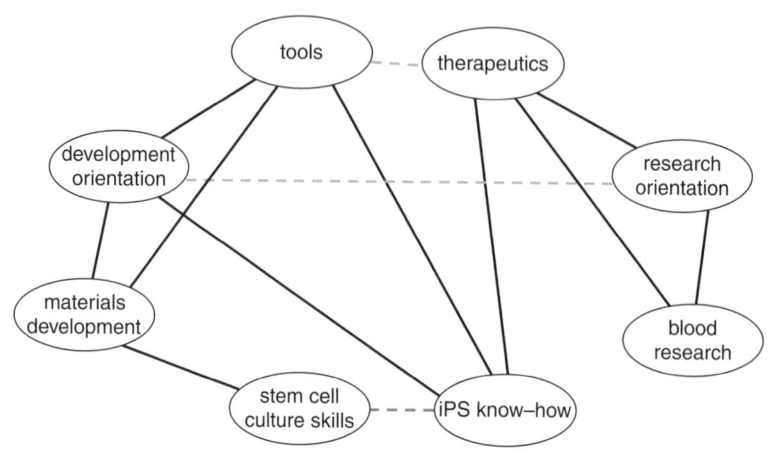

FIGURE 3.4 Elements and interactions in the new business model at CDI

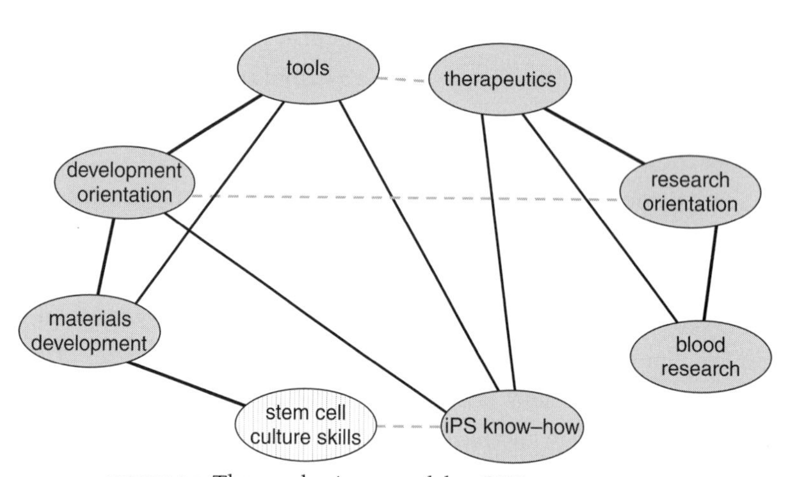

FIGURE 3.5 The new business model at CDI

Frustrated, but coherent structures

What can we learn from such an apparently abstracted exercise?

First, we note that the simulation reflects exactly what happened at CDI. Following the uptake of iPS know-how from the University of Wisconsin-Madison, executive management made the decision to merge the separate entities together. Second, we see that in some cases, perhaps especially in entrepreneurial organizations, the relative cost of operating with internal conflicts is preferable to resolving those conflicts, presumably because there is a high cost associated with that adaptation. Third, we see this descriptive outcome operates without the explicit inclusion of key operational characteristics of the organization that would otherwise seem to be relevant or even dominant – in particular financing requirements and shareholder information and preferences.

Why are these observations important?

First, strategic theories of fitness must be applied cautiously in entrepreneurial contexts. Organizations in relatively formative stages can survive, and may even need non-complementary elements. Extracting those elements decreases the organization's overall effectiveness or legitimacy, while adapting those elements to achieve fitness may require resources that are simply not available. This does not suggest that "fitness" is not an appropriate link between organizational characteristics and performative outcomes. But it does suggest that theories of strategic complementarity should be primarily applied to large, stable organizations.

Second, entrepreneurial processes may not be decomposable into component elements and activities aimed at achieving predefined configurations of strategic advantage. In other words, there are indeterminacies in entrepreneurial processes that include suboptimal choices. This is an extension of the path dependency argument, but in an entirely different direction. Theories of strategic complementarity presume a specific type of relationship among elements called supermodularity. Supermodularity means that there is a monotonically increasing function of complementarity among elements that may be achieved in step-wise fashion by swapping out, adapting, adding, or excising elements. But at entrepreneurial organizations the condition of supermodularity simply isn't met. For the most interesting entrepreneurial organizations, the ones developing truly innovative technologies, processes, or entirely new value creation systems, there may be no continuous path to fitness.

This is comparable to mountain climbing or maze solving. Nature provides no rule requiring that the route to the summit only incorporates path choices that lead upward. Alternatively, imagine a maze that has double-backs and dead ends. A heuristic of path selection based entirely on moving in the general direction of the exit could very well lead to failure.

The upshot is that despite the idiosyncrasies and complexities of real organizational systems, we can simulate business models. But doing so requires applying a narrative lens to complement traditional strategic theory. Although the challenge is apparently dramatically more complicated and non-obvious than previously supposed in strategy, it is still possible to develop viable approximations of organizational behavior *even without accounting for every seemingly important component.* Unlike strict strategic logic, narrative coherence allows for vagueness, overlap of elements, and the complexities of interpretation. For example, the concepts of therapeutic and tool development elements in the CDI example incorporate certain assumptions about resource intensity and risk/reward profiles that are then reflected in how managers interpret the conflicting interaction between them. The flexible structure allows the model to account for highly sophisticated interactions in relatively simple but robust heuristics.

THE FALSE PROMISE OF COMPLEXITY

With few exceptions,[9] the research and practice of entrepreneurial management and corporate strategies have become more complex over time. This is, in fact, a general trend in scientific knowledge creation and understanding. As observational tools, computational capabilities, and information integration modes become more sophisticated, it becomes theoretically possible to apply ever-increasingly complex models to explain observations. At the extremes, supercomputers and distributed computing clusters labor to solve or approximate massively complex computational problems in science and engineering. An excellent example is NASA's Pleiades, one of

Table 3.3 *Why complexity analysis fails for entrepreneurship*

	Physical (weather)	Strategy (competitive positioning)	Entrepreneurship (opportunity identification)
Granularity of data	High	Moderate/ High	Low
Fidelity of data	High	High	Low/Moderate
Objectivity of communication	High	Moderate	Low/Moderate
Established standards of expertise	High	High	Low
Fidelity of conjoined assessment	High	Moderate	Low
Continuity	High	Moderate/ High	Low

the most powerful supercomputers in the world, incorporating more than 85,000 processor cores and capable of more than one thousand trillion operations per second. Pleiades has been used for a variety of NASA mission analyses, including aerodynamic models that assess the threat to spacecraft posed by debris shed during launch.[10] Tremendous benefits accrue from the utilization of massively complex systems, including more accurate weather forecasting.

But there are important distinctions between physical systems, corporate strategic decision-making, and entrepreneurial contexts. These distinctions have implications for the application of complex modeling and the potential power of simplicity, as shown in Table 3.3.

The available data for physical systems is often highly granular, incorporating standardized measures, such as air pressure and humidity. At the corporate strategy level, some data can be very granular, such as market share and prices, but much data is less detailed or unitized, such as a firm's extant or potential asset stock or tacit capabilities in a given skill or task. At the level of opportunity identification, all of the critical data may be of low granularity, such as future market need characteristics.

The fidelity of available data varies similarly. Physical systems may yield multiple measures of testable data, while strategic systems may present measures that can't be easily tested, such as the efficacy of a firm's learning routines. At the level of opportunity identification, however, data fidelity may be low or even effectively zero. For example, no reliable models exist to predict rates of life science innovation. Despite billions of dollars of research over decades, we have no way to predict or even estimate a likely timeframe for the development of cures for numerous diseases and conditions including Alzheimer's, cancer, and psoriasis.

Entrepreneurial opportunities are far more difficult to communicate effectively than strategic concepts or physical system characteristics. Established standards of expertise in novel technologies or emerging customer characteristics are similarly limited.

Perhaps the most telling indicators are challenges to conjoined assessment and continuity. Fidelity of conjoined assessment measures whether the combined efforts of acknowledged experts are likely to lead to convergence. In general, physical systems benefit from multiple assessments, as expertise tends to be additive. Although this broad framework has come under scrutiny in the field of strategy, business schools teach strategy development as a tractable problem that benefits from experience, wisdom, and conjoined analysis. But entrepreneurial action is once again different – it is an unpredictable and highly individualized discovery process. Combining high uncertainty and low fidelity analyses does not, generally speaking, improve the quality of the analysis.

Finally, there is the challenge of continuity. Physical systems are subject to physical continuity – despite Hume's philosophical arguments that the temporal uniformity of nature cannot be proven. In other words, science relies on the assumption that physical laws remain constant in time. Physical systems will present the same outcomes tomorrow as they do today. And, on the whole, corporate strategy contains a presumption of continuity as well. Past successful strategic processes and configurations may be assimilated and thus applied to future examples. But entrepreneurial activities, especially opportunity identification, enactment, and exploitation, may be inherently unknowable. They rely in part on changing the very nature of the competitive landscape. In other words, even if opportunity identification worked in one way previously, it might not work the same way again.

Increasing the complexity of an analytical system to improve the accuracy of outcomes only works if the inherent error rates associated with additional measures are low. And if the number of variables is high, it only requires small inaccuracies in the system, or a couple of highly inaccurate measures, to invalidate the process. In other words, making a predictive system for entrepreneurial outcomes is, for now, difficult or impossible, both because the measures and fidelity are poor and outcomes lack continuity.

THE FALSE PROMISE OF SIMPLICITY

For all these concerns, the neural network model, which ignored a lot of details in the organizational circumstances at CDI, still predicted the structural change. Is the solution to complexity to narrow down to a simple set of the most important factors? Unfortunately, this is, itself, an oversimplification. Let's see why.

To solve purely mathematical problems, we address the number of parameters, the measurement error for those parameters, and the strength of the interaction between parameters. But rather than delve into the mathematical mechanics of complex systems, let's consider a fairly simple example that most people have struggled to resolve in one form or another: the dinner seating arrangement.

It's very likely that you or someone you know has had to put together a seating arrangement for a wedding or other significant event. It seems quite simple: a set number of people need to be assigned to seats at a relatively fixed number of tables. It's just a question of whom to put where.

But human systems have intricacies and complexities that are both instantly obvious to identify and not obvious to solve. Aunt Lydia with Cousin Katherine can't be seated together, as they aren't talking to each other. And best not to put Uncle Richard with Daniel, as they do tend to drink a bit too much and end up singing inappropriate songs. And it would be best to put as many of the teenagers together at one table or they'll just be sullen and miserable with their parents, but Sarah and Katherine shouldn't be seated together as they both dated that one boy ... Wouldn't it be nice to put Daniel with Sarah, though? They'd be so cute together. Perhaps putting all of Elizabeth's friends at one table, or at least near each other? And so on.

In theory, the seating chart problem can be optimized by developing a set of critical characteristics for individuals, assigning categorical or numerical values to those characteristics, and then creating some sort of mathematical function that describes the system based on the combinations of characteristics of individuals at each table. Then all that information gets packaged off to NASA's Pleiades computer and out pops the best solution.

Realistically, for a problem with, say, 100 people, ten tables, and ten characteristics, a supercomputer wouldn't be required. A desktop and a programmable spreadsheet application would probably be sufficient. Of course, there is the minor challenge of identifying the ten key characteristics, assigning values to those characteristics for each person, and then cross-checking to make sure that the characteristic profiles are consistent and valid.

No-one we know solves the seating problem this way – not even our most hardcore engineering and programming friends. Everyone we talked to solves this problem in precisely the same way: assigning people to tables somewhat at random at the start, with reference to a limited set of the most obvious heuristics, and then sorting it out as we go along, adding in heuristics that arise based on observations of the emerging configuration.

We accomplish this based on some very primitive rules, along the lines of:

- Start with a couple of "important" or "obvious" people that need to be seated in specific places
- try to create groups of people that share key characteristics (teenagers)
- don't put people together who really dislike each other (Aunt Lydia and Cousin Katherine)
- accept smaller conflicts to resolve bigger ones (can't put Daniel with Sarah because Sarah really should sit with her father, Richard).

But now we recognize these rules. They are not rules of strategic complementarity or mutual enhancement – they are rules of coherence. In other words, coherence provides a viable and commonly used method for solving what are otherwise complicated or intractable problems. In fact, coherence may be a dramatically better alternative for a variety of reasons:

- High cost of data collection or analysis
- Low benefits associated with minor improvements due to dynamic effects or difficult-to-measure outcomes
- Significant uncertainty in data fidelity or measurement of effects
- Uncertainty in continuity of data or effects over time

And now, of course, we recognize these conditions as precisely those we associate with entrepreneurial action.

THE PROBLEM WITH DETERMINISM

There is one more problem with simplicity that needs to be mentioned. Often the heuristics that become important only arise because of the configuration that emerges. Small changes in the initial seating assignments might have drawn attention to entirely different heuristics. For example, seating families together, rather than grouping the teenagers, wouldn't have required delving into teenager dating history, but might have involved considering generational conflicts.

In other words, the development of coherent solutions is itself path dependent, precisely because most human and organizational systems present multiple viable configurations. That is, initial configurations are important, because sense-making depends heavily on attention, which drives the first stages in the solution-seeking process.

Not just simple rules or competencies

Various strategic theories have been developed that could be mistaken for coherence. Because of their prevalence in the practice community, we compare entrepreneurial coherence to two of these: simple rules and core competence.

The strategic theory of simple rules was developed by the eminent organizational scholars Kathleen Eisenhardt and Donald Sull. Eisenhardt and Sull have studied some of the problems with traditional strategic management. They considered the challenges of rapid accelerating data availability, variation in data reliability, uncertain and shifting competitive contexts, and high opportunity costs for delay. They studied some of the most successful growth companies of the 1990s, such as Yahoo!, Autodesk, Cemex, and AES. The theory of simple rules suggests that these firms succeeded precisely because they focused on following a few central heuristics rather than complex and sophisticated strategic plans. Examples of these heuristics include: "split a business when it reaches more than $1 billion in revenues" and "only acquire companies with less than 75 employees."

Hindsight shows that the success of some of the exemplary firms highlighted in their research was short-lived or illusory. Yahoo! lost more than 80 percent of its market value in the dot-com crash. The success of some firms in the study, such as Enron and GlobalCrossing, turned out to be fundamentally illusory. While we should not judge the strength of the simple rules framework on these counterexamples alone, we might choose to see simple rules for what they are – circumstantial codifications of narrative coherence.

In one sense, following simple rules is itself a relatively simplified heuristic for implementing coherence. Simple rules implicitly allow managers to overlook minor inconsistencies to the benefit of the system as a whole. In a fast-moving technology sector like network infrastructure and security, firms like Cisco can't afford to spend a year on due diligence for each new acquisition target. The rewards for successfully integrating leading-edge technologies can be extremely high, and the cost of missing a key technology is similarly high. On the other hand, the cost of overpaying for such a technology or buying a few technologies that don't pan out is relatively low. Focusing on rapid, small acquisitions in target areas fits the narrative of leading-edge growth.

On the other hand, simple rules must be, by definition, context specific. Cisco has executed more than 100 acquisitions for which price data are available, and the vast majority fit the simple rules mantra, as shown in Figure 3.6. But what is perhaps striking about this data is the fact that the sum of payments for the 107 smaller acquisitions ($18.8 billion) is less than half the amount paid for the 11 larger acquisitions ($38.8 billion). It's also interesting, if perhaps obvious, to note that the year in which Cisco spent the most on acquisitions was 1999 (> $16.5 billion), a year in which Cisco's market value peaked above $500 billion and overpaying for acquisitions was a reality of the market.

The problem, then, is that simple rules work when they work, but they have to be ignored when they don't work. We know that economies, industries, and competitive and technological contexts may change dramatically. If that's the case, what happens to the simple rules? Clearly they must change. They change to account for and solidify the new coherent configurations. Coherence is the driving force behind simple rules, and incoherence determines when simple rules will have to be updated.

Not just core competence

One of the most powerful and long-standing strategic frameworks is the theory of core competence popularized by Gary Hamel and C. K. Prahalad. An outgrowth of the resource-based view of the firm, core

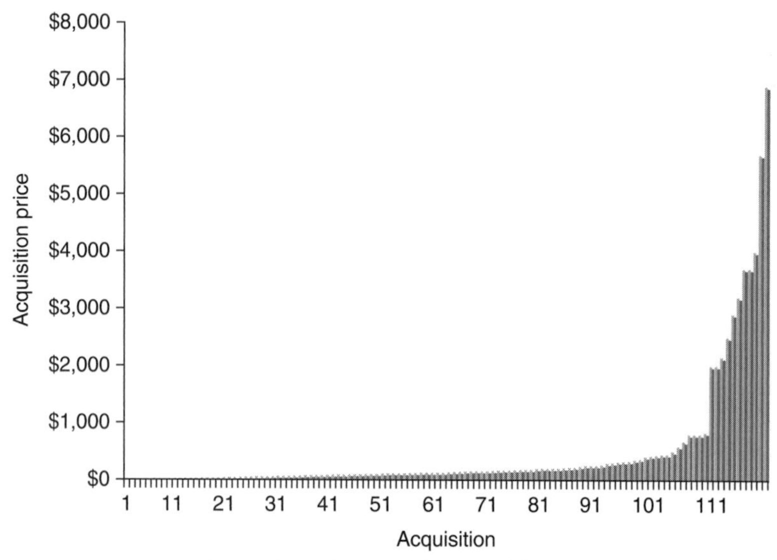

FIGURE 3.6 Acquisitions by Cisco (arranged by size $m)
Source: Cisco acquisition data aggregated at www.en.wikipedia.org/wiki/List_of_acquisitions_by_Cisco_Systems

competence builds on established principles of strategic positioning. First, firms compete in relatively well-defined arenas. Second, firms have limited but potentially unique and valuable resources. Finally, firms that efficiently leverage unique resources are inherently better positioned to create and capture value.

Core competence is also an overspecification of coherence. While simple rules ignore the dynamic perspective of coherence, core competence excises complexity in favor of certainty. In other words, rather than partial overlook of inconsistencies, core competence recommends eliminating the inconsistent elements in the system. The problem with this approach is that the option to overlook partial incoherence allows organizations more flexibility and opportunity-seeking behavior. Strategically, core competence may increase a firm's probability of effectively competing within a specific competitive context. It is unlikely, however, to be the best option for entrepreneurial or large firms seeking novel opportunities or undertaking fundamental organizational innovation. Our own research using the IBM data set showed that simplifying structures by simply extracting non-core activities did not facilitate strategic flexibility. Firms achieved flexibility to pursue novel opportunities by simplifying structures without giving up control of non-core functions. Firms that pursue core competence do achieve coherence, but they do so in a way that isolates the firm's core value creation functions and increases rigidity.

CDI – REMODELED FOR COHERENCE

Entrepreneurship, then, isn't just about excavating diamonds from a mine. And it isn't just about making things that no-one else has thought of. Entrepreneurship is about shaping opportunities with the most powerful tool available: the organization. It is the process work and storytelling enacted by the organization that brings the most interesting, industry-changing opportunities to reality. Our third insight, *remodel the organization for coherence*, builds on the simple premise that opportunities are imperfect. When entrepreneurs realize that opportunities may be shaped, adapting the organization to that purpose becomes the critical enabling process. Remodeling the organization and the narrative to ensure coherence makes the opportunity accessible. This is fundamentally a sense-making process that happens concurrently with the operations of the business. It is important for entrepreneurs to understand that this sense-making will take place whether they participate in it actively or not, at every level of the organization. Harnessing that narrative change and evolution towards the firm's goals facilitates a heuristic of simplicity without sacrificing the flexibility to adapt as opportunities change.

Nowhere in our study was this more evident than at CDI. The newly combined company focused on the development of a single assay cell type, cardiomyocytes, and subsequently developed a high-throughput process for the manufacturing of that cell type for drug discovery purposes. The manager of corporate development, who previously had managed a staff of two, was given authority to ramp up sales and marketing activities. By early 2010 the external-facing function within the organization had ten full-time employees. Throughout the organization, employees noted a shift, described as "research to development," or "development to manufacturing and sales." The interviews conducted in early 2009 displayed predominantly relaxed, optimistic, and cheerful tones. The interviews conducted in late 2009 suggested higher stress levels and tensions, sometimes explicitly linked to the restructuring, sometimes linked to the new focus on commercialization. In general, mid-level employees appeared to be the most affected by the transition, both in terms of changing responsibilities and change in attitude or outlook.

It was impossible to ignore the internal conflicts presented in the new configuration. But the simulation showed configuration to be the coherent, stable solution. How did that translate into operational changes and outcomes?

The new configuration of the organization placed primary importance on manufacturing, sales, distribution, and support processes. At the same time, significant research activities continued, including project scoping and selection activities worth noting. For example, the firm used an internal "call for projects" activity in late 2009 to identify high-potential new product areas. Management winnowed ten proposed projects to three and tasked inter-functional groups with preliminary research to demonstrate feasibility. Final proposals were presented in February 2010 and a single project chosen for further funding.

And, in fact, CDI appears to be gaining momentum. In August 2009, the company added Leroy Hood and George Church to its Scientific Advisory Board. Dr. Hood and Dr. Church are two of the world's most celebrated scientists in genomics and have been instrumental in the formation and success of more than 20 successful biotechnology companies, including Amgen, Applied Biosystems, Millipore, and others. In December 2009, CDI announced the launch of iCell™ cardiomyocytes for drug development testing. In April 2010 the firm closed on $40.6 million in venture financing, bringing total venture funding to more than $70 million.

Coherence is not perfection. But for CDI, it made sense.

NOTES

1 See, for example, "The trouble with lab-created stem cells – and why they won't displace embryonic ones" by Andrew Moseman for *Discover* Magazine Online, February 16, 2010.
2 See "Stem cell therapy in China draws foreign patients" by Louisa Lim as reported on National Public Radio, March 18, 2008.
3 Sarasvathy, S. 2001. Causation and effectuation: towards a theoretical shift from economic inevitability to entrepreneurial contingency. *Academy of Management Review*, 26: 243–288; Downing, S. 2005. The social construction of entrepreneurship: narrative and dramatic processes in the coproduction of organizations and identities. *Entrepreneurship Theory & Practice*, 29: 185–204.
4 Lounsbury, M. and Glynn, M. A. 2001. Cultural entrepreneurship: stories, legitimacy, and the acquisition of resources. *Strategic Management Journal*, 22: 545–564.
5 Fisher, W. R. 1987. *Human Communication as Narration: Toward a Philosophy of Reason, Value, and Action*. Columbia: University of South Carolina Press.
6 www.classics.mit.edu/Plutarch/theseus.html (accessed January 15, 2011).
7 Anand, V., Ashforth, B. E., and Joshi, M. 2005. Business as usual: the acceptance and perpetuation of corruption in organizations. *Academy of Management Executive*, 19(4): 9–23.
8 Zahra, S. A., Sapienza, H. J., and Davidsson, P. 2006. Entrepreneurship and dynamic capabilities: a review, model and research agenda. *Journal of Management Studies*, 43(4): 917–955.
9 Eisenhardt, K. M. and Sull, D. 2001. Strategy as simple rules. *Harvard Business Review*, 79(1): 106–116.
10 Source: www.nas.nasa.gov/Resources/Systems/pleiades.html.

4 Build bridges

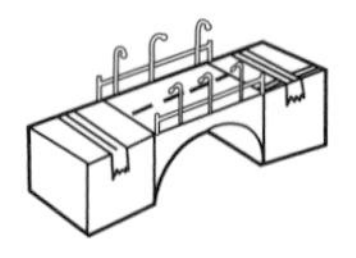

THE VOXEL.NET CONNECTION

On the top floors of a historic building in lower Manhattan, an implausible opportunity is being shaped. This firm operates on the 30th floor, among the dense cluster of skyscrapers that reaches from the grid of New York streets to the clouds. Voxel.net (Voxel) is itself stretching for the clouds, in a very literal sense.

In 1996, Raj Dutt was an undergraduate student at Rensselaer Polytechnic Institute (RPI) in New York. But his real passion was leading-edge network systems, and by his third year he was running a server capacity brokerage business when he wasn't in class. Eventually he teamed up with Zachary Smith, a Juilliard School alumnus, and they began building a company. This company's success has come via an atypical route.

Voxel evolved without venture funding. It isn't leveraged. It never moved far from its origins in New York, arguably one of the most expensive places on Earth to operate. Considering that the company's expertise lies in software engineering and network administration, it is rather amazing that it has never outsourced employment, despite access to low-cost capabilities in Asia enabled by the very technologies that Voxel propagates. A picture of the senior management team on the company's website looks more like a college social club than a New York technology firm.

But Voxel grew from humble beginnings as a server capacity broker to owning its own servers and selling value-added capacity management services. Now it does something that would defy most business school strategy course lessons. With less than 100 employees, Voxel provides global cloud-computing services, including a fully integrated content delivery network. And it competes head-to-head with companies like Google, Amazon, Intel, and Rackspace, whose combined market capitalization exceeds $400 billion including more than $40 billion in cash assets.

This small company serves firms that need a high performance web presence delivering high bandwidth data and content. Incredibly, Voxel provides high-touch service by linking customers that need help directly to the engineers who administer their network solutions. Voxel manages to do this at comparable or lower costs than firms more than a thousand times bigger. In other words, Voxel provides a differentiated product yet competes on price without scale economies. This is strategically non-obvious. After all, in technology-intensive, high-velocity markets, surely the advantage goes to the leaders:

> The industry leaders [in cloud services] in the public and private worlds have a very large chunk of the market, and catching up is a daunting prospect. Unlike other industries, there is just as much raw innovation coming out of the market leaders as there is out of the upstarts. Such is the challenge, even though the market itself is relatively nascent and evolving very quickly.

That's not a Gartner report or someone from Amazon – it's Raj Dutt blogging about the reality of the industry.[1] Even Voxel's CEO seems to appreciate the strategic implausibility of success. But, in 2009 Gartner named Voxel as one of its most innovative "Cool Vendors," and in 2010 Voxel was identified as one of the fastest North American networks for cloud service, beating out companies like IBM.

There's one more piece of information worth knowing – Voxel relies on open-source software and solutions. Although it has developed a significant amount of software and network administration tools on its own, it commonly shares them with customers and competitors alike. Want to find out how the company implements its core products? Download its 40-page technical guide from the website. How do Smith and Dutt make this work? And how have they consistently bridged the gaps in technology and product development to bring this business so far? Can they keep on doing it as they go from 100 to 1000 employees or more?

INSIGHT 4: BUILD BRIDGES TO SPAN OPPORTUNITIES

Some opportunities are long-lived, even by organizational standards. Union Pacific has been hauling goods, especially coal, across the United States for more than 150 years. Some Japanese firms can trace ancestry for a thousand years. But even longevity is no guarantee of success: the 250-year history of *Encyclopaedia Britannica* was threatened when it was slow to move from print to digital format.

By comparison, Voxel is still in its infancy. In only ten years, however, Voxel has probably built, transitioned, and rebuilt more business models than Union Pacific and Encyclopedia Britannica combined. The information technology space evolves hundreds of times faster than railroads and educational content, though the effects of that rapid evolution have been felt by companies in lower-velocity industries.

To be sure, Voxel has been nimble, flexible, and opportunistic. But it has also gone further than just agility. It has helped shape the development of opportunities by operating at the leading edge. Sometimes it has stretched beyond the leading edge of commercially viable technology. For example, Voxel launched the VoxCAST content delivery network (CDN) before industry leaders Rackspace and Amazon brought comparable offerings to the market.

At each crossroad, at each chasm, Voxel has leveraged its resources to become something greater, something unexpected. Voxel's executive partnership of Raj Dutt and Zachary Smith is characterized by consummate bridge-building. The team, and by extension the company, has spanned numerous opportunities by redesigning organization and even reshaping the nature of the opportunities addressed. It has overcome many of the liabilities faced by small companies to compete on an international scale, and has even used the irreversibility of some investment decisions to its advantage. As the landscape changed, it either changed with it or even anticipated it.

But past adaptability does not guarantee future survival. Voxel faces incredible challenges to remain competitive, and the traditional management literature provides little guidance or hope for success. It must overcome significant disadvantages. Low-scale operations without price premiums limit the firm's return potential. Scaling faster than well-financed competitors would require relatively massive investments. Neither seems likely. Then again, the traditional management literature wouldn't have predicted its success so far; Voxel is a strategic anomaly. Voxel's success hasn't been about strategic positioning, it has been about the experiential process of bridging opportunities.

ENTREPRENEURSHIP AS EXPERIENCE

Earlier, we focused on the nature of opportunities. We examined how entrepreneurs perceive opportunities and the dynamic relationship between entrepreneurship and opportunity realization. In an expanding, accelerating entrepreneurial context, it is time to update frameworks for understanding entrepreneurial cognition. The nature of the industrial world has changed dramatically, facilitated by changes in technology, especially information technology. Traditional models of entrepreneurship incorporate theories of human behavior that were focused primarily on explaining large-scale phenomena. These theories of rational action made sense when scale economies drove most industrial outcomes. The generalizations required by economic and even social models of entrepreneurship were imperfect, but provided the most convincing explanations of entrepreneurial outcomes.[2]

We know, now, that while these remain effective descriptors at the macro-level, they are imprecise and possibly misleading at the level of actual individuals and early stage firms. In this new environment, local effects matter because information technology facilitates and propagates micro-scale events and knowledge globally and almost instantaneously. We expect to see more and more examples of entrepreneurial action that are difficult or impossible to explain with traditional economic tools. Voxel appears to be one of these examples.

What actually happens as entrepreneurs enact and implement novel opportunities? The first three insights describe how entrepreneurs link organizations to extant opportunities. Insights 4, 5, and 6 help explain how entrepreneurs go beyond what they already know to evolve organizations to enact incipient opportunities. We have therefore reached the halfway point on the narrative journey towards understanding the new entrepreneurial world.

To appreciate how entrepreneurs build bridges, we begin by examining the context of change in which entrepreneurs operate. We identify the design tools entrepreneurs wield to seek coherent and effective organizational configurations. These frustrated, yet flexible systems enable responsiveness and an almost intuitive appreciation for nascent opportunities.[3] We'll generalize those adaptations based on the scale of discontinuity that the firm must address.

Insight 5 will present unique capabilities and actions that distinguish the innovative entrepreneur from a strategic executive. The sixth and final insight will bring us back to the role of the entrepreneur in the rapidly evolving global context. We'll be able to show how the journey transforms the entrepreneur and simultaneously empowers the entrepreneur to achieve the unexpected.

ONE-WAY FILTERS

New companies face many challenges. Savage Entertainment emerged from the shelter of parent firm Activision to face the competitive world of video game development. In doing so, it left behind the infrastructure and stability provided by a large firm with multiple revenue streams. A variety of frailties are associated with new firms compared to old firms. The "liability of newness"[4] is commonly associated with the firm's requirement to learn new capabilities, establish effective routines among individuals with limited prior interaction, and build revenues without the legitimacy lent by prior customers.

Let's look at another example. C5–6 Technologies Inc. (C5–6) is a biofuels development company in the US. Like Savage, C5–6, C5–6 was spun out of a parent organization, a company named Lucigen. Both Savage and C5–6 benefited from the transfer of certain capabilities from the parent firm and limited legitimacy from that origination. But there were also important differences. Savage's legitimacy stemmed both from the prior experience of the management team as well as contract work for Activision in the spin-off process. In the case of C5–6, however, a more explicit link was established, both because the Lucigen shareholders were the founding shareholders in C5–6, and because the firms shared certain facilities and project work. From a knowledge perspective, the capabilities imported by Savage were relatively general and could be applied to a variety of opportunities and markets. C5–6 had access to specific patents, but that portfolio presented a relatively narrow foundation of intellectual property associated with developing biomass fuel solutions based on proprietary enzymes licensed from Lucigen.

All of these limitations represent path dependencies. Path dependence limits the options available to the organization based on prior decisions or circumstances. Path dependence is the shadow of the past, in figurative terms. Prior decisions both determine the options available to the firm as well as the sequence of actions taken by the firm. Over time, the weight of implemented decisions may gain momentum, further influencing choices despite the presence of contra-indications. Path dependence is quite common in businesses. It is a mentality that says that a product line that has been a bottom-line miracle worker receives more resources and managerial bandwidth even when there are some indications that the future might not be as bright. By now, managers are knee-deep in the product and find it sticky to wade out.

But path dependence is only half of the story. The other half is irreversibility. *Irreversibility* reflects the difficulty of reversing key decisions. Whereas path dependence reflects the nature of decision-making outcomes, irreversibility reflects decision outcomes that are difficult or impossible to rescind. Despite the apparent flexibility of many entrepreneurial firms, rapid recovery from failure is not always an option.

Many firm choices, even at theoretically flexible entrepreneurial firms, are effectively irreversible. For example, when Confederate Motorcycles accepted venture capital in an effort to scale production, a series of organizational changes were triggered that ultimately resulted in bankruptcy. The chain of events led from the growth requirement associated with investor expectations to a brief period of lower-quality production in an attempt to achieve scale. Because that initiative conflicted with founder Matt Chambers' fundamental beliefs, the Board fired him, setting in motion the downward spiral of an organization deprived of a key leader. Whether or not Confederate could have avoided those outcomes is at least partly debatable, but once it had accepted the venture capital funding – and spent some or all of it – the firm had no near-term option to reverse the event. The operational mode that was the coherent solution prior to the funding event, was no longer accessible for the organizational system. Confederate couldn't return the investment and it couldn't operate as if it hadn't received it.

The combination of path dependency and irreversibility engender a phenomenon we refer to as "one-way filters." A one-way filter is an engineering term referring to a device that enables flow in one direction but not the other. Examples of one-way filters include flow valves that pass a liquid, logic gates that pass information, and subway turnstiles that pass commuters.

In reality, most organizations pass through a variety of one-way filters on a regular basis. Product releases, strategic acquisitions, and employee hiring and firing are reasonably common business decisions that may serve as one-way filters. But one-way filters have distinct and important implications in the enactment of novel opportunities. Remember Insight 3: Remodel the organization for coherence? When large organizations pass through filters, the effects tend to be relatively diffuse and incremental, because the organizational system is already configured to operate within a competitive context. But entrepreneurial firms in development stages function in a meaning-making context that builds on internal, rather than external validation. One-way filters for these companies tend to be significant and discontinuous, and often generate unexpected outcomes.

To see this in more detail, we need to examine the three characteristics of one-way filters from the perspective of the organization rather than how it operates in the environment. The key characteristics of one-way filters are: asset specificity, learning thresholds, and network embeddedness.

Asset specificity is an established factor in a variety of strategic and operational decision-making frameworks including transaction cost economics and resource-based theories of competitive advantage. Resources acquired that are specific to a given transaction or operational capability may be difficult to utilize for other activities or opportunities. When transaction or value-specific resources are not fungible, those transactions or activities demonstrate high asset specificity. High asset specificity means that a resource has a limited, if important, use, and may not be extended to create value by other means. Entrepreneurs, especially innovative entrepreneurs, address asset specificity regularly. Entrepreneurs often rely on bricolage, the ability to generate value from otherwise value-less resources[5], or effectuation, the process of defining goals based on available resources.[6] Entrepreneurs who are designing or remodeling firms for coherence, or building bridges to new opportunities, utilize emergent processes for resource acquisition. In these cases, high asset specificity limits avenues of entrepreneurial action. The acquisition of transaction-specific resources may be necessary, but often proves irreversible.

Learning within organizations is similarly an important element of competitive advantage and firm success.[7] Entrepreneurs often learn in real-time without the benefits of slack resources to separate learning from implementation.[8] High *learning thresholds* associated with complex and tacit capabilities or extremely specialized knowledge incur high investment costs. Even when learning thresholds do not incorporate asset specificity, entrepreneurs investing to reach learning thresholds bear significant resource and opportunity costs. In addition, high learning thresholds often result in high affinity for technology, team, and the opportunity itself, which decreases the entrepreneur's ability to make effective long-term decisions.[9]

Finally, an entrepreneur's networks may play a key role in firm formation, knowledge and resource acquisition, and growth.[10] *Network embeddedness*, however, may come at a steep cost.[11] Network connections are generally based on relationship investments, both for initiation and maintenance. There are two implications to incurring high network costs when entrepreneurs exploit opportunities. The first is simply the opportunity cost associated with the investment in the network, which may limit the entrepreneur's development of other opportunity-specific resources. The second is the effective investment in a narrative specific to the firm's trajectory that the entrepreneur propagates within the network to engender legitimacy and acquire resources.[12] Narratives within networks are double-edged swords. If the narrative is incoherent or misapplied, the firm may be unable to access the resources it requires. If the narrative is too convincing and "sticky," the entrepreneur may find it impossible to deconstruct or detangle that narrative as opportunities evolve or emerge that require bridge-building. Table 4.1 summarizes the path dependence and irreversibility components of each component of one-way filters.

Table 4.1 *Components and determinants of one-way filters*

	Path dependence	Irreversibility
Asset specificity	High asset specificity commits the organization to specific resource structures. Business model innovation requires simplifying structures, which may then require the firm to divest apparently valuable but opportunity-limiting resources. Specialized competitors often require high asset specificity to distinguish the firm's value creation structure from other businesses.	More rarely, high asset specificity commitments are irreversible, because they are linked to multiple business model structures. Venture capital investment, for example, underpins both resource structure and value creation structure, because it necessitates specific value creation priorities.
Learning threshold	Achieving learning thresholds creates path dependencies by embedding competencies within specific business model elements or structures. This is essential for more innovative firms to routinize value creation for efficient deployment. Some industries and technologies present a narrow range between the learning threshold and the leading edge. In these environments, firms that choose to compete on competencies may be required to invest continuously in those competencies at the risk of missing other opportunities.	Achieving learning thresholds in rapidly shifting environments exposes many innovative firms to competency traps. These irreversible filters lock in competencies even after the market requires new capabilities. Firms achieve the unexpected when they successfully gamble on irreversible learning thresholds that engender scale effects or lock in customers.

Table 4.1 *(cont.)*

	Path dependence	Irreversibility
Network embeddedness	Even more than assets and knowledge, networks are sticky. Embedding the firm within a network or ecosystem is usually necessary for legitimization. Narrative legitimacy may be difficult to alter, creating path dependencies based on the very competencies required for initial success. Proximity to high density network hubs, while valuable for technology propagation, also tends to generate narrative inertia.	Network embeddedness is irreversible when firm narratives become identified or enmeshed with the network narrative. Firms that achieve the unexpected draw on the network narrative without becoming reliant on it.

There is a one-way filter through which all organizations pass. The forming of an organization is a unique and critical process. *Imprinting* is a phenomenon in which organizations acquire characteristics associated with environmental context during formation. Many organizations retain those founding characteristics, sometimes longer than necessary. But our interest in imprinting is not limited to the structural characteristics of the entity; we want to understand the sense-making story that converges at the organizational level, whether driven by the entrepreneur or not, in the pursuit of the opportunity.

Opportunities change; the pursuit of opportunities changes the entrepreneur. The imprinting process is therefore important to understand to appreciate entrepreneurial action. Consider Savage. For more than a decade Sonny and Morten's company provided contract development services to major video game studios. Yet less than six months before the company's business model became effectively non-viable, Sonny, Morten, and most employees saw the company as something other than what it was. They wanted to believe, and thus still did believe that, at heart, Savage was a creative content developer rather than a job shop. The founders' original goal of building an independent creative studio had survived despite ten years of contrary evidence in the form of the firm's own activities.

WHEN LANDSCAPES CHANGE

One-way filters are an important reality for entrepreneurial firms, especially firms exploring new spaces and new customer needs. But the effects of one-way filters, especially when those filters are irreversible, become even more powerful when entrepreneurial firms operate in changing landscapes.

Opportunity landscapes and competitive landscapes are two different, but partly related, concepts. An opportunity landscape represents the spectrum of unmet and unknown opportunities available for entrepreneurs to exploit. As we've stated previously, however, we believe that a static representation of opportunities is misleading, as the action of entrepreneurs both reveals and shapes opportunities. Regardless, it is sometimes helpful to envision opportunities in a spatial arrangement to demonstrate that some opportunities are similar while others are dramatically different. For example, Skype and WebX are different businesses providing different services to different customer types, but the underlying nature of the opportunities present important similarities. Both were enabled by the massive expansion of low-cost or free internet bandwidth as well as the necessary network protocols and software to enable audio/video transmission online at very low cost. We may argue about how "close" Skype and WebX opportunities are; they certainly have more in common with each other than cardiomyocyte-based drug assays or $85,000 carbon-fiber motorcycles.

The metaphor of the landscape is also helpful because we often describe entrepreneurs as seekers or explorers who survey the environment and identify opportunities based on prior experience, knowledge, or interest. We may imagine that some opportunities are near or far, based on the resources or time required to access them by a specific entrepreneur. Opportunities present larger or smaller value creation potential, and greater or lesser risk of failure.

A competitive landscape is a similar metaphor that describes the cohort of competitor firms within an industry. This more traditional framework focuses on understanding the nature of competition and the relationship among firms within an industry. Again, we might differentiate landscape position by relatedness of products, services, or capabilities, or entirely different criteria such as organization size, profitability, or geographical location.

These metaphors lend themselves to a discussion about changes in landscape. Every organization faces the probability or certainty of ongoing change, whether in the industry context, competitive interplay, or market profile. Landscape changes, whether faced by entrepreneurs or strategic competitors, may be seen as local or global, slow or rapid. We continue the landscape metaphor by describing the four primary types of landscape change: earthquakes, sinkholes, glaciation, and erosion. These are not specific to entrepreneurial firms, of course, but successful response and adaptation in the face of these different challenges requires varying amounts of entrepreneurial action. We focus, of course, on the effects of landscape change on entrepreneurial firms, and in particular how innovative firms survive and sometimes even thrive on these unpredictable events.

Rapid, global landscape change: Earthquakes

Major changes in competitive landscapes are generally easy to identify, at least in hindsight. The launch of iTunes provides an excellent example. Only a year after launch, Apple had sold about 100 million songs over iTunes and sales growth was accelerating. As noted previously, iTunes revenues will surpass all music CD sales in 2011 or 2012, a watershed of sorts for the dominance of digital media distribution. One year in the life of an entrepreneurial firm may seem like eternity, but one year in the life cycle of an industry technology is a relatively short span. In some industries the accelerating pace of technological change demands even faster product launches, but the advent of even truly remarkable new products or services, such as eBay or cellular phones, much less medical technologies like stem cells, tends to develop over years rather than weeks or months.

These rapid, global landscape changes are, in effect, earthquakes. They include product launches like iTunes, discontinuous supply or demand shocks such as the 1973 oil crisis or the liquidity crisis of 2008. More rarely, they include technological advances, but radical innovation does not always equate to change magnitude. It is not the innovation, but the entrepreneurial opportunity that determines the speed of adoption and impact on related industries. Massive, one-time shifts in the competitive context may create dramatic repercussions that can be traced directly to a single marketplace event. Napster was launched in June 1999, and was sued by the RIAA less than six months later. Lawsuits by major recording artists Metallica, Dr. Dre, and Madonna followed in 2000. Napster shut down operations in 2001, only two years after launch. In the interim, estimates suggest that Napster users were downloading more than 2 billion songs per month, almost one thousand times the number of iTunes song downloads per month in 2010. The effect of peer-to-peer networks and related innovations presented a challenge to the recording industry on a scale the studios simply hadn't imagined. The technology was only one enabler; the extraordinary scale of uptake was driven, in part, by high bandwidth and protective privacy policies at universities. This combination set the stage for the opportunity associated with digital media distribution.

The advent of music downloading was an earthquake. The impact was felt in distribution channels and beyond. Musicland was an American chain of retail music stores that experienced rapid expansion in the 1980s and 1990s after acquisition by Sam Goody. By 2001, when the Sam Goody business was bought by BestBuy, there were 1300 Musicland stores in the US. This was, of course, right at the time that internet usage and peer-to-peer networks enabled illegal downloading of music, and not even the RIAA's successful lawsuits that shut down Napster could slow the demise of Sam Goody and the Musicland stores. The stores were sold to Sun Capital in 2005, and went into bankruptcy in 2006. By 2007 only about 100 stores were still trading, and all were converted to sell other retail products. But the effects were felt in industries further afield. Music producers benefited from the shift in power in the supply chain. And developers of electronic music, including hobbyists, obtained an almost zero-cost propagation tool. A variety of software packages emerged to support legitimate and illegitimate music and video copying. As in most cases of destruction in opportunity landscapes, creativity filled the gaps.

C5–6 Technologies experienced a similar event. In the late 2000s, the ongoing rise in crude oil prices helped push investors towards technological alternatives to petroleum. In 2007 and 2008, a variety of circumstances caused prices to rise more than 100 percent. Investments in alternative fuel technologies and businesses spiked as well, as investors and entrepreneurs wondered if oil prices would reach $200/barrel and possibly continue upwards. In the midst of this run-up, C5–6 was making big plans. The company negotiated a joint venture with an international biofuels company, moved forward on novel enzymatic treatment processes, entered into talks with some of the largest biofuels stock materials companies in the world, and made pitches to Silicon Valley venture capitalists.

Then the bottom fell out of the oil market, almost overnight. Oil prices plunged because of the recession. C5–6 Technologies' enzymatic solutions to improve biofuel efficiency became less economically compelling. Financing for ventures in biofuels dried up worldwide, and C5–6 had to carefully assess its options. In the end, it became clear that commercial success was further off than hoped.

Rapid, local landscape change: Sinkholes

Sometimes rapid change happens in a highly localized context. Sinkholes can be very destructive to a localized set of organizations or a single value chain. Common examples are changes in local regulations or the economic ripple effects induced by an event at a larger local organization. Many cities in the United States have enacted anti-smoking legislation. While the broader social good of these measures is defensible, specific businesses such as bars and taverns may have been adversely affected. The argument has been made that because many individuals smoke when they drink, they are less likely to frequent bars if smoking isn't allowed. Anecdotal and some quantitative evidence suggests that bars where anti-smoking legislation has been passed have seen one-time drops in business. It isn't clear, however, if these represent long-term changes in which smokers visit bars less or find alternate locations where smoking is still allowed, a far less common option nowadays.

Sinkholes appear to be relatively rare, and highly specific. They are often trailing-edge events from larger trends. Social and economic trends are, in effect, analog – that is, they progress in small increments. But organizations are, in effect, binary – an organization is ongoing or it has failed. The cumulative effects of change, such as the failure of American auto manufacturers to develop competitive operations and cost structures compared to Japanese manufacturers was a key component of the bankruptcy of GM and Chrysler during the economic recession of 2008–2009. The decisions to shutter specific plants, while triggered by longer-term trends, were sinkholes for many businesses in the local communities where those facilities were based.

A sinkhole can, of course, be a local natural disaster, such as the arrival of Hurricane Katrina in New Orleans. The destruction of Confederate's manufacturing facility and nearly all their paper records and information systems, as described in the next chapter, is an excellent example.

Slow, global landscape change: Glaciation

Of all the landscape changes entrepreneurs face, the slow, global change of glaciation is perhaps the most difficult to deal with. Earthquakes are violent, obvious events that happen quickly, often capturing critical managerial attention. Glaciation, on the other hand, happens slowly and insidiously, requiring much more acumen and perceptivity. Glaciation reflects large-scale but gradual change in industry dynamics, economic context, factor markets, or competitive capabilities that may be difficult to detect via short snapshots. And, even when some aspects of change are clearly detectable, the implications may not be obvious or even relevant until the change reaches a key milestone or another change in the environment or industry suddenly makes that factor relevant.

Savage Entertainment provides the best example of the effect of glaciation from our case study companies. As noted, the cost of video game production has risen steadily over the past decade. Industry participants have been aware of the change, but the implications were not fully appreciated until two other factors came into play. The first was that larger video game developers increased in-house capabilities to accommodate the increasing complexity of development. For most studios, completely outsourcing development generated too many risks. A minimum level of development capacity in-house was necessary both as a backstop for meeting deadlines and to ensure that games fit with creative vision and downstream marketing and distribution. That capacity was often filled by acquiring smaller studies. These two glacial changes collided mid-way through the financial crisis of 2008–2009. Because game development often takes two years or longer from scoping to release, the effects of the crisis in the game industry lagged reduced investment decisions at the production studios. Video game revenues continued to rise, but major studios scaled back on new projects, anticipating the need to limit R&D expenditures. This allowed them to source nearly all new development work in-house, accommodated by capacity increases. These factors came together in 2009, when the studios effectively met their demand for development with internal resources. Firms like Savage found that there was no contract work to be done. The slow, but perhaps inevitable glacier arrived, carving out the landscape from beneath their feet.

Slow, local landscape change: Erosion

Erosion, like sinkholes, tends to be local, but can have larger downstream effects. One example of erosion in our study group is ValueLabs' need for additional space amid the explosive growth of technology businesses in Hyderabad. ValueLabs found itself desperately in need of high-quality space for its growth cadre of software programming teams supporting global IT and dot-com companies. Land development and construction couldn't keep up with business growth, and the costs of facilities had increased so rapidly that renting space had become more expensive than building. The timeframe required to obtain all the necessary government permits, design, and build facilities had become too long for firms to wait.

ValueLabs was unable to make the necessary investments because local conditions for new infrastructure development had eroded. Costs were high and government regulations and involvement in facility procurement had not kept pace. This led to ValueLabs' decision to slow growth rather than pursue less ethical means of obtaining the necessary access for facility development.

CHANGING LANDSCAPES

Businesses worldwide face accelerating technological evolution, broader competition, and unexpected economic and social changes. Given that all firms, small and large, regularly address novel opportunities, we want to be able to move beyond small business examples and context. Smaller firms' characteristics and actions are more contingent on local contexts and the imprinting effects of founding. How do larger firms address landscape change and implement business model innovation? This is an important question, because some small firms will become large firms. Are coherence and bridge-building relevant outside the purely entrepreneurial firm? Do problems of scale, structural rigidities, and relative resource munificence change the game?

In order to get a high-level perspective on how firms respond to changing landscapes, and in particular how CEOs sculpt organizations and interpret the outcomes, we need to look at larger organizations innovating to pursue opportunities.

As mentioned previously, we worked with IBM's Institute for Business Value (IBV) to examine this issue. IBM's 2006 Global CEO Survey provided data from interviews with more than 750 CEOs of leading organizations worldwide. We specifically focused on companies initiating business model innovation. These are organizations focused on innovation associated with the identification and pursuit of new opportunities to expand the enterprise. Business model innovation may look different at large firms than in small firms, in part because small firms' business models tend to be only partly formed. The change processes may also look different, because the structures at large organizations may be more well-formed, hierarchical, and concrete than at smaller firms.

But the results of the detailed analysis were surprising. The mechanisms are different, but the underlying processes employed and the effects on activities, resources, and interactions match what happens at entrepreneurial firms addressing new opportunities. This is an important finding, because it means that large firms may learn from the opportunity strategies of smaller firms.

We have already noted that firms enacting business models are often responding to large-scale changes in the industrial or competitive environment. Of course, what seems like large scale to a small company might not be as noticeable at a large company. Large firms tend to address earthquakes and glaciation rather than sinkholes and erosion because the scale of operations often spans local regions. In the end, however, when large firms innovate on their business models, they have a similar goal in mind to small firms. That goal is to identify opportunities and maintain or improve the firm's strategic flexibility to address those opportunities.

WHEN BUSINESS MODEL INNOVATION MATTERS

An important lesson about business models as levers for shaping and exploiting opportunities is that large firms need to learn and apply this language, perhaps even more than small firms. Highly innovative entrepreneurs apply some of these tools intuitively, because narrative rationality is fundamental to individual cognition and behavior. But many small firms have a highly centralized locus of control, often spearheaded by the founding entrepreneur. This is generally not the case with larger firms, especially larger firms that are one or more generations removed from the founders and key growth managers.

These firms then have a greater challenge: they must remain competitive within their industries, as well as find ways to survive and evolve in the new entrepreneurially driven world. Much of the twentieth century was characterized by the evolution of competitive strategy, and especially how those strategies reflected and were reflected on organizational structures. The most successful firms have relied on *systematic innovation*, incorporating ongoing, cyclic, regenerative, and accumulative innovation processes.[13] But in the past 25 years, a new generation of powerhouse organizations rose from nothing to global prominence, often by exploiting or creating entirely novel market opportunities. Google is considered one of the most powerful companies in the world, and the company has begun to challenge established giants in sectors, industries, and markets with an entrepreneurial logic unlike anything the Fortune 500 has seen before. Firms as diverse as publishers, media companies, telecommunication firms both in infrastructure and consumer devices, venture capitalists, energy utilities, and data storage companies have cause for concern with each press release from Menlo Park.

In the next 25 years, the challenges for large firms will only increase. Millions of entrepreneurs around the world are gaining access to the information and networks of a more and more democratized landscape of opportunities. Large firms will face more hurdles leveraging prior knowledge from successful innovations, because the very nature of innovation is changing, as our research below shows.

IBM's insight

Many larger corporations have begun responding to the challenge.

A few years ago, IBM saw some of the early signs of this transformation. In its Global CEO Survey, it added a section on business model innovation to complement the traditional areas of technology, product, and process innovation. IBM's unique position as the world's largest information technology company gave it unparalleled access to the senior leadership of the most powerful and dynamic companies on the planet. The results of the survey were unequivocal, though we have yet to extract the full depth of its lessons.

The first lessons are straightforward – although less than a quarter of the organizations participating in the survey identified business model innovation as their primary innovation focus, the outperforming firms were the ones that focused on it (see Figure 4.1). And those business model innovators were outperforming their cohorts within each industry (see Figure 4.2) Business model innovators were more likely to rely on CEO leadership for innovation than firms targeting product or process innovation. These were also actively developing complex networks of collaborators to assist in the process of seeking out new opportunities and developing new business models.[14]

IBM generously provided access to the underlying study data. We extended IBM's original analysis to uncover some of the key characteristics and processes of business model innovation.

"Big" business model innovators

Business model innovation doesn't look like product, technology, or process innovation. Whereas traditional innovation efforts are driven by market and competitive forces, large firms initiate business model innovation in response to much broader, even more complex factors (see Table 4.2). Globalization, environmental pressures, and sociopolitical changes are leading large organizations to reassess the opportunity landscape. This confirmed our expectations that larger firms approached business model innovation, in part to respond to the global landscape changes, earthquakes and glaciation. These forward-thinking firms have realized that process and product innovations, while necessary for maintaining near-term competitive positions, won't be sufficient to compete with firms emerging from unexpected and unfamiliar places.

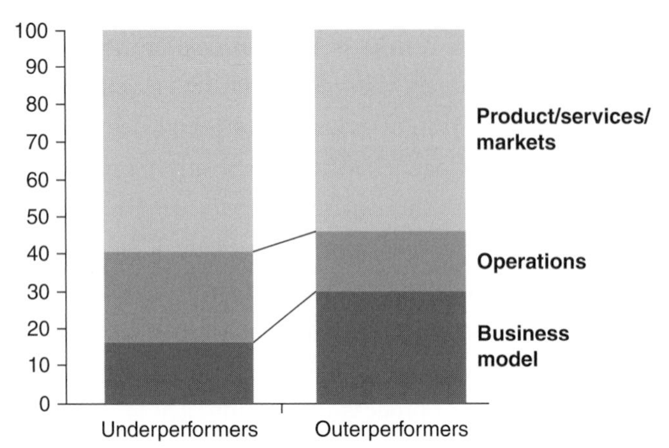

FIGURE 4.1 Outperformers focus on business model innovation
Source: Giesen, E., Berman, S. J., Bell, R., and Blitz, A. 2007. *Paths to Success: Three Ways to Innovate Your Business Model*. Published by IBM Institute for Business Value. Document Number G510–6630–01.

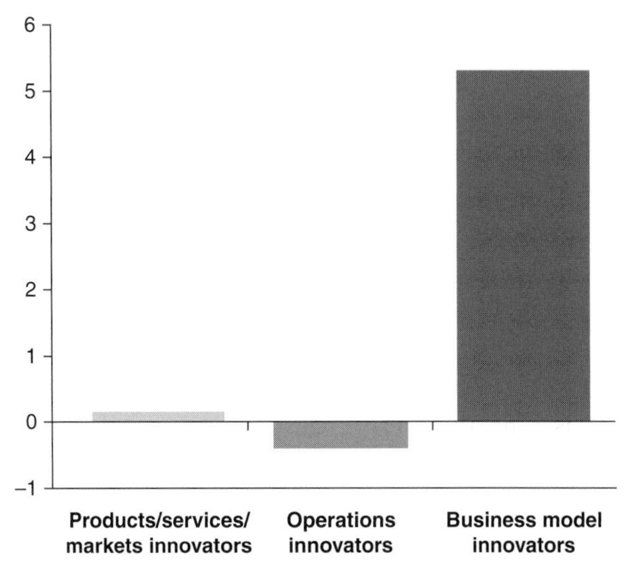

FIGURE 4.2 Business model innovators outperform cohorts in operating margin growth
Source: Giesen, E., Berman, S. J., Bell, R., and Blitz, A. 2007. *Paths to Success: Three Ways to Innovate Your Business Model*. Published by IBM Institute for Business Value. Document Number G510–6630–01.

Table 4.2 *What drives business model innovation at large firms?*

Drivers linked to business model innovation at large firms	Drivers not linked to business model innovation at large firms
Macro-changes	Prior change success
CEO leadership	Perceived magnitude of change requirement
Non- EU/global firm operations	Industry or nation-base

This change in perspective represents an important shift in how large firms approach the fundamental question of long-term success. Traditional strategy focuses on competitive positioning against known competitors in established markets. Some research has approached strategy from the perspective of targeting new markets, but even these frameworks utilize a strategic mindset in which the firm anticipates leveraging extant resources into related market segments. There is some evidence that firms have natural capabilities to develop an intuition of nearby opportunities.[15] In the new entrepreneurial context, these nearby opportunities will be accessible to a wider spectrum of firms and entrepreneurs. The firms in IBM's survey achieving the best results were the ones who were looking beyond current markets and current competitors for glimpses of the unknown. These organizations have moved beyond *optimizing strategy* into the realm of *realizing opportunities*.

Aside from responding to a different set of forces, business model innovators must approach opportunity with new tools and new thinking. One of the most important facilitators of business model innovation is a firm's adaptive capacity. How quickly and effectively can a company redirect resources and rewire connections to address new opportunities? Research has shown that these *strategically flexible* firms succeed in turbulent and hypercompetitive environments. But how do firms implementing business model innovation maintain or grow this capability?

The Global CEO survey provides important clues.

When the landscape changes ...

The data showed that large firms make specific choices to maintain and build strategic flexibility. Some of the results are straightforward. Companies should maintain a creative culture which smoothes innovation discontinuities and cushions change efforts by reducing internal inertia and active resistance. But as suggested by our research on how small companies build, design changes are just as important. It is the design changes that present real challenges for innovative organizations, because some of the effects are not apparent or intuitive.

There are two key factors in making the necessary structural changes: focus and control.

In order to identify, assess, and exploit new opportunities, managers at firms must focus on the *external* environment rather than attempt to extend internal resources. Novel opportunities emerging from macro-level trends can't be seen within the organization. This presents obvious problems, because the vast majority of firms are hard-wired for internal observation and assessment. When companies evaluate the environment, they tend to focus on markets they already know. This is, in fact, another example of the power of cognitive fluency, where lack of familiarity creates a perception of distance. After all, highly paid managers assessing opportunities partly or completely unrelated to the firm's core business are pure overhead, contributing nothing to the firm's ongoing returns.

One strategy for improving focus is to reconfigure activities to improve operational efficiency. But our analysis of the IBM data showed that firms which reconfigured activities and structures actually reduced the likelihood of achieving reduced strategic flexibility by 30 percent, controlling for several factors such as culture, size, geography, and industry sector, among others. The age of re-engineering the corporation to exploit emergent opportunities may already be over!

Many firms have opportunities to improve efficiencies and uncover hidden capabilities within the firm. But these processes maintain or even increase inward-focused managerial attention. Such activities may generate short- and long-term benefits to the firm, but they do not reduce structural complexity. In the end these firms become less strategically flexible, less able to meet the challenge of new opportunities.

The data shows that reducing, not reconfiguring structural complexity is essential. Business model innovators must find a way to decrease inward-focused attention. The data from these leading organizations shows they accomplish this by decreasing the complexity of organizational structures. Business model innovators reduce structural complexity by delegating responsibility for non-core activities to other organizations. This is, of course, a long-standing practice, and may appear consistent with theories of core competence or systems of strategic complementarity. But, as is often the case, the reality is not that simple.

Focusing on core competence is effective advice to develop and maintain advantages in a firm's primary business area. This is an inherently internally oriented strategic philosophy. Focus on firm strengths and don't get distracted by issues or opportunities beyond that sphere. Do you recall our path dependence discussion? Unfortunately, this is precisely the type of path dependence we have noted. This will work well for firms in stable, low turbulence industries, especially when those firms are comfortable in primarily defensive roles. But the emergent entrepreneurial environment will not provide firms, especially large firms, the luxury of focused competence. The timeframes in which focused competencies are dominant and sufficient are becoming shorter.

For example, Nokia had become the world's largest cell phone manufacturer based on sophisticated miniaturization technologies that enabled smaller, lighter, more reliable phones sold at a premium price. This advantage lasted less than five years. But it wasn't eliminated by competition from other cell phone makers competing on similar criteria. The innovation that transformed the phone user experience shifted adoption criteria from size, telecoms functionality, and fashion, to entertainment and a broader user experience. The iPhone arrived, leveraging the iTunes Store and an open innovation model that completely changed the market criteria for what a mobile phone should do. The iPhone provided a variety of telecoms-related functions, such as GPS and web surfing. But it was a thousand other things as well: a level for hanging pictures, a gaming platform, even a tool for identifying popular music titles from a few hummed bars. Nokia's core competencies simply weren't relevant when the business model for successful mobile technology changed. By late 2010, the dominant players in the cell phone market were Apple and Google, neither one a mobile phone company by nature. But they realized that the hardware had become commoditized. They expanded their horizons to identify value creation in software, user experience, and open innovation. We believe similar transformations are coming in many other industries and product groups. Focusing on and even leveraging core competencies won't be sufficient to ensure that organizations succeed when the game changes.

The surprising corollary to this has to do with partnering. Most strategy research demonstrates the power of collaboration. Access to more knowledge and network effects complement and leverage assets and capabilities. Effective collaborators build on their existing advantages or even partner their way out of trouble. Extensive studies of social networking and shared learning environments demonstrate that companies big and small benefit from trusted relationships for accessing information, importing competencies, and bringing new products to market. Agreed!

But in the new world of entrepreneurial opportunity and business model innovation, these *strategic* lessons lose some of their sheen. The firms in the IBM study that relied on partners for business model innovation reduced their likelihood of achieving strategic flexibility by 40 percent (see Figure 4.3)!

In other words, business model innovators get more benefit from reducing complexity than interaction with partners. It

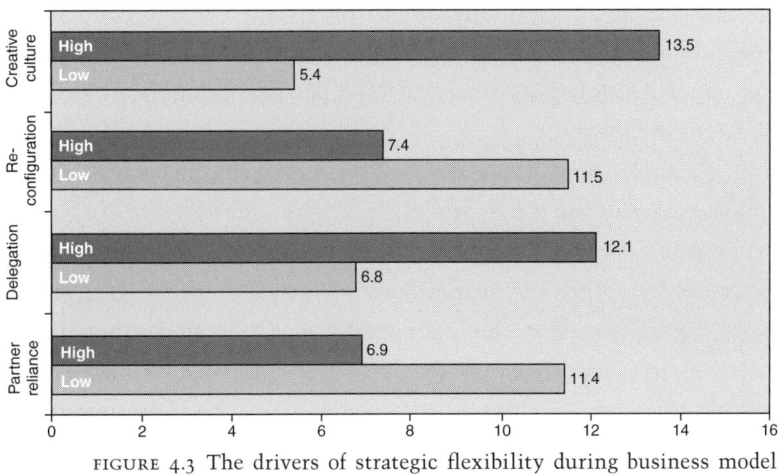

FIGURE 4.3 The drivers of strategic flexibility during business model innovation

is possible that this is an artifact of the nascent state of business model innovation. After all, most firms are not aware of the new requirements for long-term organizational innovation. Even fewer have internalized the capabilities to consistently implement business model innovation. In fact, the IBM data showed that prior success change was not a significant determinant of whether firms enacted business model innovation in preference to other types of innovation. Perhaps in another decade or two companies will have the entrepreneurial skills to more effectively support opportunity-seeking efforts with partner companies without the threat of missing opportunities themselves. The data could look very different in 25 years, but for now, reducing complexity trumps working with partners.

The real story, however, is even more complicated. First, there appear to be fundamental costs associated with reliance on partnering that outweigh the strategic flexibility benefits during business model innovation. Firms perceive opportunity landscapes differently, so collaborating on business model innovation requires sophisticated and costly coordination activities. Although collaboration is extremely effective at leveraging synergies and extending proven capabilities, it actually appears to handicap firms implementing business model innovation, precisely because the novelty of expanding opportunity sets is likely to take partnering firms in entirely different directions.

Second, there are nuances to simplifying structures when innovating the business model. It might seem logical and even prudent to take delegation of activities and structures to its logical extreme, by spinning off non-core activities and divesting unrelated businesses. But the unexpected result is that when firms ceded control of those apparently non-essential areas, the potential gains to strategic flexibility disappeared. The innovator firms that delegated, but held control, improved the likelihood of achieving strategic flexibility by 50 percent. It appears that control is as important as focus. One possible driver for this outcome is knowledge spill-ins, which transfer important, validated heuristics across organizational structures.[16] While these must be constantly reinterpreted and re-validated, the coherence-seeking value of these heuristics may be lost when non-core functions are severed entirely.

In this new world, firms need to delegate non-core activities to empower core competencies and to retrain critical managerial focus on the external environment. This helps identify opportunities that would not be obvious, or even logical, framed against the firm's established capabilities. Organizations must also retain control even as they reduce complexity. This ensures access to information about the broader environment and leverages a more diverse portfolio of capabilities towards new opportunities.

The simple lessons are illuminating. When the landscape changes, simplify design to survey opportunities – and if you can – change landscapes by building bridges to pursue new opportunities. And don't rely on partners to help you build those bridges – the odds are they are busy building their own bridges to entirely different opportunities.

INFLECTION POINTS

Understanding the importance of one-way filters, changing landscapes, and business model innovation brings us back to the role of the entrepreneur as bridge-builder. Even nascent, partly formed organizations have already been affected by one-way filters, especially when landscapes shift. These inflection points may represent windows of opportunity. In these moments, the firm may overcome inherent limitations of current opportunities and firm structures. Alternatively, firms may fall victim to landscape changes when entrepreneurs are unable to design and execute the structures necessary to achieve new and unexpected value.

Regardless of the landscape change, there are three transitions that entrepreneurs access, enable, or construct from scratch to reach new opportunities. They are crossroads, bridges, and skyhooks.

Crossroads

Entrepreneurs, and managers more generally, tend to envision decisions as crossroads – the choice of one from numerous available options. The analogy is often quite apt: such decisions progress the firm along a path dependent route in which prior resource investment decisions directly impact the perceived availability of options.

C5–6 Technologies found itself at such crossroads in 2009. Spun out of another small biotechnology firm, the company was primarily committed to the commercialization of a specific enzymatic technology platform towards increasing biofuel production efficiency. But the company was struggling to prioritize previously funded research against uncertain but potentially high-value commercial opportunities. If the firm focused on grant funded research, much of which was tightly linked to the parent firm, it would benefit from the stability and certainty of that funding and the interaction with the larger organization. But it was unclear whether that process would lead to near-term commercial applications. On the other hand, the company had received expressions of interest from a variety of potential biofuel partners who wanted to see an advanced proof of concept or pilot scale test before investing. Pursuing this presented numerous challenges, as private financing had all but dried up amid the global financial crisis, and even the successful development of a pilot test was no guarantee of future partnering or investment. In fact, although the company had already established one joint venture agreement with a large international biofuels company, problems had arisen in the funding from that partnership and the outcomes of the development process were in doubt.

From a business model perspective, C5–6 had numerous frustrated structural elements. On the one hand, it was tightly tied to the technology, ownership base, and funding mechanisms of the parent company. At the same time, it had brought in an outside CEO specifically to try to explore the opportunity space for growth options and new commercial applications. The extant grant funding, much of which was linked to the parent organization, provided an important cash source for both organizations, but was arguably less critical to the long-term success of C5–6. Finally, the lead scientist at C5–6 was also closely associated with the parent organization, and served in a role with affinities for both firms.

In the end, the crossroads proved a decisive time for C5–6. Despite the recommendation of the CEO, the Board chose to focus on the safer cash flow stream of the grant funded research. The departure of the CEO, on good terms, may have been a direct effect of this. C5–6 continues to make progress developing its unique technology. It reached a crossroads and selected a path from the available options. Only time will tell the outcomes of that choice.

Bridges

For some organizations, arriving at crossroads presents multiple unattractive options. Sometimes organizations attempt radical business model change to build a bridge to entirely new opportunities.

This was the case for Savage Entertainment and Sustainable-Spaces. We've already talked about Savage's situation, but it is useful to clarify the efforts it made to build bridges, even though its efforts were ultimately unsuccessful. As the full scale of the environmental change became clear, founders Sonny and Morten were faced with difficult choices. They knew that they didn't have the cash to support their full staff without new contracts, but they also knew that releasing staff would make it more difficult to close deals. This was especially relevant because once they released their designers and developers those individuals would be significantly less amenable to returning, weakening their bargaining position with potential partners or acquirers.

The founders had spent years talking about developing original content, and their proprietary development engine offered some unique capabilities. But when they sat down to discuss how quickly they could put together a new proposal they realized that the scale of development costs dramatically exceeded their available resources. They contacted outside experts about the prospects for raising private funds, and they also approached larger developers about acquisition scenarios. All of these bridges were incomplete, in part because they didn't address the fundamental business model flaws exposed by the changes in the industrial and economic context. These bridges weren't inspirational because they didn't address opportunities both unique to Savage and novel enough to be investable. Interestingly, the founders did consider transitioning the organization to outsourced development of social networking and mobile applications. In our interviews with the founders, we could see that their hearts were not in this particular bridge. The founders of Savage were diehard gamers who'd lived the dream, first at Activision and then at their own firm. As Sonny once commented: "My job is to bring to life giant fighting robots. What more could I ask for?" Even if they could have constructed a bridge from giant fighting robots to simple social networking applications like Farmville, it is unclear whether Sonny and Morten were willing to cross it.

Of course, it is easy to criticize failure, and much more difficult to prove that an alternate course of action would have worked. Perhaps adjacent but more disparate opportunities could have been tapped, such as the defense contracting projects Savage worked on. These types of projects often have extremely long lead times and had not generally been as lucrative for the company. Perhaps there was a bridge to non-video game technology applications, such as a "build your own giant fighting robot" website. Perhaps, just perhaps, there were no viable bridges.

Savage lived, though in a greatly reduced form. Morten has a few programmers working on small outsourced projects for former clients. But Sonny has moved on. Perhaps the best indicator that there may not have been any viable bridges is exemplified by his choice. The landscape change was dramatic; it is possible that all paths led in only one direction. As of December 2010, Sonny followed the path of least resistance. He went back to work for Activision, where the story started more than ten years ago.

SustainableSpaces was altogether different. Matt Golden and his team had become national leaders in residential energy efficiency auditing, remodeling, and political action. Matt seemed to be spending as much time in civic speeches and testimony before legislative bodies as he was doing home audits and perfecting the audit tools. But he could already see that the footprint of his work was limited. He could only audit so many houses. He could only train so many other people to audit houses at the level of quality and professionalism required. Auditing and remodeling residential properties was an inherently local business. There were too many local weather, construction practice, landscaping, and zoning characteristics to easily transfer all the necessary information beyond a modest radius. Matt wasn't happy with this outcome – his goal was large-scale residential energy efficiency improvements. But he didn't have the right business model.

Matt began building a bridge. He talked with his team about creating a larger footprint. He thought about sustainability, and the mission of the company. He discussed his interests and managerial limits with his team. He talked about finding a way to dramatically expand the scope of the company; to facilitate rapid expansion of energy efficiency auditing. He described the effect that could have on energy utilization in the United States. He assessed what that would mean for the company, for American families, and for global ecology.

He began tapping his network of connections. He didn't know exactly where the bridge was going, but he knew he needed someone to help him build it. He was not a private finance, operational scale-up, or product management expert. He began meeting with financiers and retired executives to talk about his vision for growing the business. He described bridging the company's current opportunity for local energy auditing to a national agenda for saving energy, money, and the environment.

A venture capital connection led Matt to Pratap Mukherjee, who had recently sold his roll-up services business and was looking for a new challenge. Pratap wasn't as driven by the environmental mantra as Matt, but he understood and appreciated those values. He saw an opportunity to contribute via his direct experience with operational scale-up. Mukherjee and Golden spent numerous hours discussing the bridge that SustainableSpaces would have to build. It would not be a simple or a short bridge. They agreed that the current opportunity simply could not be scaled up. They decided to construct a bridge to an opportunity in which neither had any experience. They decided to transform SustainableSpaces from an energy auditing and remodeling company into a software business that served other construction firms.

Constructing this bridge was no simple feat. First they had to raise venture capital to completely transform the current organization. That's not something venture capitalists normally are very keen on. VCs prefer to invest in the scale-up itself, once the investment seems likely to generate near-term returns. But Mukherjee and Golden found venture partners prepared to cross the bridge with them on the theory that it was the only bridge available to address the opportunity afforded by scaling energy efficiency auditing.

They had to completely restructure the firm's business model, including nearly every formal and informal structure at the organization. The resource structure of the firm needed to be adapted to a high investment, low variable production cost model. The transactive structure, though not immediately required, would need to focus on sales to businesses rather than sales to homeowners. Post-sale support and product maintenance would be the service model, rather than quality operational effectiveness. The entire value structure of the organization would have to be turned completely upside down. Rather than focus on idiosyncratic value at each home, the company needed to find a way to routinize the auditing process, formulize remodeling recommendations, and present information in a way that educated construction companies rather than homeowners.

The new company, Recurve, has continued to benefit from government incentives for energy audits and energy-efficient remodeling investments. But Recurve is a company that isn't relying on the government for its success. Needless to say, the evolution of the organization remains a long, somewhat shaky bridge. But the early indications are that the company has made significant progress. It released the beta version of its software in April 2010, and raised $8 million in private funding in June. It was #682 in the Inc 5000, and 24th among energy companies, posting a 450 percent revenue growth rate from 2008 to 2010.

Even if the bridge is considered complete with the launch of the beta software, the long-term success of the company is far from assured. The new challenges of running a software company with a construction business inside are significant. But without the bridge the company would still be that construction business, and that just wasn't going to meet Matt Golden's definition of success.

Table 4.3 *Crossroads, bridges, and skyhooks*

	Crossroads	Bridges	Skyhooks
Action	Choice from opportunity set	Create option by spanning opportunities	Exploit unexpected and unpredictable opportunity
Resources	Available at firm	Accessed via narrative	Retooled or gifted
Risk	Low	Moderate to High	Unknowable
Entrepreneur's role	Strategist	Author	Gambler

Skyhooks

Some companies see crossroads and choose a path. Some, like SustainableSpaces, ignore the paths and build a bridge to span an opportunity gap. Some, like Savage, may have no choices at the crossroads, whether they want to build a bridge or not. And some may not have the option to build a bridge. It's not certain, but that's our best guess for the team at Voxel.net. Voxel appears to need a skyhook.

A skyhook is literally a hook on a cable used to pull something up from its environment. Polymath scholar Daniel Dennett applied the term to refer to an aspect of design complexity that couldn't easily be explained by extrapolating from prior information, experience, or structures.[17] Although Dennett used the term to deride theories of intelligent design in the evolution debate, we use it in its more operational sense: a mechanism of change without obvious antecedent or predictability. It is, perhaps, the stroke of fortune that favors the prepared, which will later be rationalized into genius or raw determination.

Skyhooks are, of course, the rarest of events in organizational change. They cannot be planned, and probably not encouraged. They may be of the entrepreneur's doing, but more likely the instigating or enabling event comes from the outside. Table 4.3 compares the three bridging mechanisms, along with the risk-resource profile and the role of the entrepreneur.

We chose our study companies precisely because all of them exhibited unusual circumstances, behaviors, changes, or outcomes. But even within this unusual group, only a few have truly experienced skyhooks. Let's take a look at the ones that did and how they happened.

The first is Broadjam. Broadjam's skyhook was due to a purely external trigger – the launch of iTunes. iTunes was definitely not a skyhook for Apple Computer, it was the calculated result of years of effort that included security encryption research, software design and engineering, extensive market testing, and most importantly, hardball negotiations with the most unlikely of allies – the slow-moving and tradition-bound music studios. The implications were obvious to many industry observers – two days after the April 28, 2003 launch, *BusinessWeek*'s Alex Salkever wrote:

> The wraps are now off Apple's long-rumored music-download service. If anything, the April 28 launch of the iTunes Music Store was anticlimactic. Not because it didn't live up to expectations but because it so easily did. But don't be fooled. This service promises to forever alter the balance of power in musicland ... I think Jobs's maneuver will go down in history as the final straw that broke the back of the old music-distribution system ... I give credit to Jobs for seeing this opening and driving the Apple cart right through it.[18]

The launch bridged Apple's transition from a computer company to a consumer electronics experience business. It also provided a bridge to the ailing music industry, which needed a way to continue to monetize its music catalogs threatened by peer-to-peer sharing networks. The music industry would have undoubtedly found a solution eventually; the sheer value of those catalogs warranted investment; though it's possible that a few years' delay could have resulted in more industry shuffling of players and the demise of the weaker players.

For Broadjam, iTunes was a true skyhook, even if the founder, Roy Elkins, didn't fully appreciate it at the time. It was important, but the full impact on the industry and Broadjam took a little longer to emerge. First, the digital format, and more importantly, the transmission of digital files had been legitimized. This has had wide-ranging effects across the music industry, and in particular the non-consumer music industry. Previously, the music studios had significant formal and tacit control over the use of music for commercial purposes such as movie scores and television advertisements. Music producers would contact the studios and describe what they wanted, and studio representatives with an exceptional but often heavily tacit knowledge of the studios' music catalogue would come up with a few suggestions, from which the producer would make a selection. Over the next few years as the catalogs were fully digitized, the music producers effectively gained power over the studios, not just because they had access to more tracks, but because they could go outside the studios to independent musicians for professional quality music without paying studio royalties.

Enter Broadjam.

Elkins and his team had amassed more than a quarter of a million songs on their website from independent artists, that is, artists who hadn't been signed by the studios. And now commercial music producers were ready to flex their supply chain power and access those songs, but they needed a search mechanism. Broadjam already had it.

Built in part on the consulting work the company had done for one of the major studios, Broadjam had a fully functional back-end for its music catalog. The system stored characteristic information about the song (length, genre, artist information) but, more importantly, it also provided instant access to the "hook," the catchy refrain or melody that epitomized the tone and gestalt of the entire song. After four years of struggling to find the right market niche, Broadjam grabbed hold of the skyhook and began generating revenue from independent musicians who finally had a mechanism to put their music in front of producers.

The skyhook at Cellular Dynamics (CDI) was primarily technological in nature – the isolation and culturing of iPS cells[19] created new challenges, but it resolved a series of technological, legal, and potential ethical issues facing the organization. It presented a clear platform technology for the development of multiple products across market types, and facilitated a coherent plan that resonated with investors.

Voxel has, in fact, already built and crossed many bridges. Raj Dutt and Zach Smith have traversed multiple chasms: they scaled a purely virtual capacity brokering business until they could acquire their own servers. They invented value-added services to differentiate an otherwise strictly commodity server capacity business until they could acquire network resources. And then they built a truly extraordinary bridge. With almost no operating leverage, financial or human resource slack, or pre-paid customers, they found a way to extrapolate their hosting network into a fully integrated, global content delivery network. They launched it before either Amazon or Rackspace, and have since maintained world-class uptime and network speed capabilities.

In pure organizational capacity terms, this is little short of miraculous. With less than 100 employees, only half of which are software and network engineers, Voxel effectively realized the same innovations as Amazon and Rackspace with less than 1/10th of 1 percent of the resources.

Of course, there are caveats. Although Voxel has industry-recognized clients such as SourceForge and the *New York Observer*, it hasn't served truly enormous content delivery clients like Yelp! (Amazon) or the University of Chicago (Rackspace). And it never patented its technologies, though it is unclear whether they could have been patented. It's possible the patenting process would lag the technology development cycle in this particular field. Voxel has begun to find itself in an increasingly difficult industry position. It cannot market and sell at the scale of the largest players. There is a growing cadre of smaller, highly specialized niche competitors targeting specific industries, geographies, and applications for content delivery networks and cloud-based hosting.

Voxel is operating in an industry that is not experiencing traumatic shifts like earthquakes or sinkholes. The change in industry capabilities may be the clearest example of rapid glaciation today. That process is powered by numerous driving factors: accelerating availability of information technology, similarly accelerating telecommunications access, rapid growth in human skills in large, untapped human resource markets in China, India, Indonesia and elsewhere, and rapid uptake of technology across almost the entire spectrum of industries worldwide. These, in turn, are driven by cost pressure, ecological issues, and the ongoing creation of entirely new consumer needs.

Nearly every economic, organizational, and strategic success factor is working *against* Voxel. It does not appear to have the option to scale rapidly enough to compete against companies like Amazon. It is unclear whether the organization as it currently exists could survive such growth if funding were available. The founders' roles and involvement would change drastically in that scenario. Specializing into a niche provides an option of apparently limited attractiveness, partly because the company has already invested into a broad-based platform designed to serve any company, anywhere. Becoming a purely low-cost competitor would fail to monetize that investment, while upcharging for services would present an enormous change in the firm's strategic position and market-facing activities. While any of these are at least possible, none of them build on the firm's strengths or successes to date.

Building bridges is one of the key capabilities that distinguish the innovative entrepreneur from the strategic manager. These are not assured success, as our case studies demonstrate. At the same time, bridge-building may be what distinguishes the successful entrepreneurs in the rapidly evolving markets and competitive spaces that will evolve as the new generation of entrepreneurs emerges around the world.

Bridging to the unexpected at Voxel

What will happen at Voxel? As we draft this chapter in early 2011, we don't quite know. As you read this, however, the answer might be as simple as going to its website. Is it still there? What about its client and case study list? Is the photo of smiling, young managers still beaming out from the "About Us" page?

We like the people we met at Voxel and wish them well. But the identification and exploitation of entirely novel opportunities are tricky, and skyhooks are fundamentally unpredictable events. Maybe they will ultimately settle for a niche position in a crowded space where consolidation is probably inevitable because of commoditization pressures. Maybe they will find a skyhook.

Voxel has already achieved the unexpected – it implemented a business plan that defies most theories of strategy, and built bridges to keep up and even lead industry leaders. Here are Raj Dutt's thoughts from the Voxel blog:

> That market is obviously a really exciting place to be right now, and at Voxel our product strategy has been centered around helping our customers take advantage of the best mixes of the old fashioned and the new fangled; a hybrid, flexible and open approach. In order to effectively do that we've evolved over the years with the market and started looking more like a software company than a hosting company. That kind of metamorphosis

> is a common theme for hosting companies that want to stay relevant in an increasingly on-demand world. For such companies (Voxel and Rackspace included), software has gone from being an operational necessity that impacts your efficiency to being a centerpiece of strategy and part of a serious and escalating arms race. In a lot of cases, we had to build this stuff from scratch because it didn't exist.[20]

But unlike Rackspace, for example, Voxel has no margin for error staying at the leading edge and maintaining growth. It nearly tripled usage on its cloud network in 2010, but the industry is moving faster than that. Rightscale reported a ten-fold increase in usage in roughly the same timeframe. Perhaps there will be a bridge that we can't see – certainly Voxel's CDN development and rollout was a spectacular example of bridge-building. Perhaps there will be a skyhook that neither we nor the company can see at this time. And while we know from personal experience that the inherent uncertainty in the entrepreneurial environment can be nerve-wracking, it is that same uncertainty, that same unpredictability that enables entrepreneurs to defy all expectations and reshape the world.

NOTES

1 Source: www.voxel.net/blog/2010/07/openstack-open-source-cloud-computing-plumbing (accessed December 28, 2010).

2 Delmar, F., Davidsson, P., and Gartner, W. 2003. Arriving at the high growth firm. *Journal of Business Venturing*, 18: 189–216.

3 George, G. and Bock, A. J. 2010. The role of structured intuition and entrepreneurial opportunities. In: Phillips N., Griffiths, D., and Sewell, G. (eds), *Technology and Organization: Essays in Honour of Joan Woodward (Research in the Sociology of Organizations, Volume 29)*. Oxford: Emerald, pp. 277–285.

4 Stinchcombe, A. L. 1965. Social structure and organizations. In: March, J. G. (ed.), *Handbook of Organizations*. Chicago: Rand McNally & Company, pp. 142–193.

5 Baker, T. and Nelson, R. E. 2005. Creating something from nothing: resource construction through entrepreneurial bricolage. *Administrative Science Quarterly*, 50(3): 329–366.
6 Sarasvathy, S. D. 2001. Causation and effectuation: toward a theoretical shift from economic inevitability to entrepreneurial contingency. *Academy of Management Review*, 26(2): 243–263.
7 March, J. G. 1991. Exploration and exploitation in organizational learning. *Organization Science*, 2: 71–78.
8 Miner, A. S., Bassoff, P., and Moorman, C. 2001. Organizational improvisation and learning: a field study. *Administrative Science Quarterly*, 46: 304–337.
9 George, G. and Bock, A. J. 2008. *Inventing Entrepreneurs*. Saddleback, NJ: Prentice-Hall Pearson.
10 Stuart, T. and Sorenson, O. 2007. Strategic networks and entrepreneurial ventures. *Strategic Entrepreneurship Journal*, 1: 211–227.
11 Khayesi, J. and George, G. 2011. Does the socio-cultural context matter? Communal orientation and entrepreneurs' resource accumulation efforts in Africa. *Journal of Occupational and Organizational Psychology*, 84(3): 471–492.
12 Lounsbury, M. and Glynn, M. A. 2001. Cultural entrepreneurship: stories, legitimacy, and the acquisition of resources. *Strategic Management Journal*, 22(6/7): 545–564.
13 Drucker, P. 2006. *Innovation and Entrepreneurship*. Burlingame, MA: Elsevier.
14 Giesen, E., Berman, S. J., Bell, R., and Blitz, A. 2007. *Paths to Success: Three Ways to Innovate your Business Model*. Published by IBM Institute for Business Value. Document Number G510–6630–01.
15 George, G., Kotha, R., and Zheng, Y. 2008. The puzzle of insular domains: a longitudinal study of knowledge structuration and innovation in biotechnology firms. *Journal of Management Studies*, 45: 1448–1474.
16 Oldroyd, J. B. and Gulati, R. 2010. A learning perspective on intraorganizational knowledge spill-ins. *Strategic Entrepreneurship Journal*, 4(4): 356–372.
17 Dennett, D. C. 1995. *Darwin's Dangerous Idea*. New York: Simon & Schuster.

18 "Steve Jobs, pied piper of music online" by Alex Salkever, *Bloomberg Businessweek*, April 30, 2003. www.businessweek.com/technology/content/apr2003/tc20030430_9569_tc056.htm (accessed January 14, 2011).

19 Yu, J., Vodyanik, M. A., Smuga-Otto, K. *et al.* Induced pluripotent stem cell lines derived from human somatic cells. Science DOI: 10.1126/science.1151526.

20 Source: www.voxel.net/blog/2010/07/openstack-open-source-cloud-computing-plumbing (accessed January 14, 2011).

5 Inspire the narrative

AN INDUSTRIAL "ART OF REBELLION"

The sexiest motorcycles in the world aren't designed or manufactured in Italy, Japan, or Germany. They are assembled entirely by hand using purpose-built carbon-fiber and titanium parts. For the past few years they have been assembled in a nondescript one-story red-brick building in Birmingham, Alabama in the Southern United States by Confederate Motorcycles. The industrial workshop stands in stark contrast to the glamor of its customers, who include movie stars like Tom Cruise and Brad Pitt. The history of Confederate reads like a movie script, including a self-made idealist, complex financial deals, and even a Category 5 hurricane.

The apparent incongruities don't end there. Confederate makes the $85,000 B120 Wraith, which is not just the "World's Sexiest Motorcycle,"[1] it's also the land speed record holder for its class, tested on the historic Salt Flats in Utah. Yet Confederate is not a bespoke manufacturer – it is a production company. The company manufactures a limited run of one motorcycle line at a time. When it finishes the run, it never makes that cycle again. Figure 5.1 shows the four primary models the firm has produced. At top is the original Hellcat that established and then nearly ruined the firm's reputation. Below that is the iconic B120 Wraith that has influenced motorcycle design around the world. Below that is the radical P120

FIGURE 5.1 Confederate Motorcycles models
Top to bottom: Hellcat, B120 Wraith, P120 Fighter, X132 Hellcat

Fighter that featured on the cover of the 2009 Neiman-Marcus Christmas catalog. At bottom is the newest launch, the minimalist X132 Hellcat.

How many motorcycle companies field phone calls from King Abdullah of Jordan? When international football star David Beckham places an order, what company has the nerve to inform him that he'll be placed on the waiting list? And they won't provide an estimate for when the bike will be ready?

Confederate has been bankrupt. It has accepted venture funding – twice. Founder and CEO Matt Chambers invested most of his life savings into the company – twice. It is now publicly traded on the NASDAQ pink sheets following a reverse merger. It has a listed market cap of $20 million[2] – nearly twice the valuation on comparable firm Viper Powersports despite being less than one-fifth the size of Viper.

In its 15-year history, Confederate has manufactured less than 1000 motorcycles. Only 150 of the new X132 Hellcats will be produced before the line is retired. In fact, Confederate has never generated significant profits, and has appeared to survive long stretches on sheer will. But somehow Confederate continues to produce motorcycles that polarize competitors, aficionados, and customers into raving fans and loathing critics. And the epic is ongoing. At the time of this writing the company is returning to its city of founding: New Orleans. Once a grateful recipient of financial and other incentives from Birmingham and the nearby Barber Motorcycle museum, the one-time darling of Birmingham has reasserted its stubborn and rebellious spirit to return to its true home.

How does a firm like this survive? Is it just the vision and determination of the entrepreneur? Is it just the undeniable uniqueness of its signature products, which have challenged modern motorcycle design? Is it the culture of obsession that permeates the employees who design, build, and sell motorcycles they can't afford to buy?

INSIGHT 5: INSPIRE THE NARRATIVE TO SHAPE OPPORTUNITIES

The continued existence of Confederate is, perhaps, a bit of a miracle. The company is fully into its third or fourth incarnation, depending on who's doing the counting. For Chambers, however, what really matters is the painstaking production of a unique motorcycle experience. To do this, in defiance of reasonable and rational business theory, Chambers continually inspires a narrative that surpasses the products, the people, and the profits. Matt Chambers personifies how Insight 5 helps entrepreneurs achieve the unexpected. Throughout the improbable life of Confederate Motorcycles, Chambers has woven a story that simply absorbs every event, good or bad, and bends it to his purpose, reshaping the very nature of this niche opportunity. Every setback, every triumph, every transition has been co-opted into the legend. The power of narrative not only connects, but reinterprets the connection between a small industrial manufacturing team in the American South and its wealthy, image-oriented globetrotting customers. Chambers has inspired a narrative that achieves the unexpected by turning dissonance into coherence:

> Resolution ... requires the right question, the right answer, and the right strategy. We must purely and honestly express our most sincere dreams of what should be but never was ... We must design in revolt with absolute resistance, eternally.[3]

As we discussed in Chapter 4, turbulence, uncertainty, and change often provoke organizations to target new opportunities. The IBM data showed that innovative large companies use the first four insights to restructure the organization. Simplifying structures, while retaining control of functions, focuses managerial attention towards new opportunities. These firms remain flexible enough to identify and exploit new opportunities, and achieve better competitive and financial outcomes.

But small, entrepreneurial businesses, especially in niche specializations or with limited geographical reach, may struggle to access necessary resources, even in the smaller world context. In these cases, the unique narrative skills of entrepreneurs like Chambers accomplish the unexpected in one of two ways. Sometimes they bridge opportunities by rewriting the organizational story in real time. This may change the nature of the team, the activities and even the role of the company in its industry. But, sometimes the entrepreneur rewrites the narrative of the opportunity itself, reshaping it into a preferred configuration, making it uniquely accessible to the entrepreneur.[4]

WRITING ENTREPRENEURIAL STORIES

Global social and technological changes accrued over the last 20 years have created a new context for opportunity exploitation. These changes, however, do not just facilitate an entirely new and expansive domain for entrepreneurial venturing. This evolution has dramatically increased the competitive pressures and organizational strains that new ventures face.

Some of the firms we studied live in rarified industry sectors with previously unseen rates of technological innovation. In these high clockspeed industries, learning effects may create benefits, rather than penalties for the most advanced companies. Voxel, for example, has benefited from a leading-edge technology position, though commoditization has already begun to set in. Return Path, on the other hand, seems to have a longer runway for building a potentially insurmountable position in the "safe email" world. This is no guarantee for success, but it does reflect an important distinction in business model structures that reflect industry dynamics.

Voxel's coherent business model matched leading-edge resources with information-rich client transactions and a value creation structure in a pre-scale context. But now scale effects have changed value creation structures in the industry. Voxel is being forced into a niche position, because the scaling effects in online storage are cumulative, rather than linear. The very nature of cloud computing creates capacity management advantages that increase with scale. Voxel's story may become less coherent if it tries to compete with fully scaled competitors on the terms dictated by those competitors. In contrast, Return Path's business model *builds* value as it is extended to global scale. It is important to note that Voxel and Return Path are not on opposite ends of the same problem. Cloud-computing scale primarily enables cost spreading and efficiencies, encouraging price competition. But scale in whitelisting email senders creates unique value for customers inaccessible to other competitors. This should enable price premiums.

This helps demonstrate that business models preconfigure strategic options. Although 30 years of research has generally supported the economic argument that industries provide space for multiple types of strategies, the structural characteristics of viable business models within those industries make some strategies inherently more difficult, and potentially less rewarding than others.

Equally important, the viability of business models, as well as the profitability of industries and the strategic specializations within those industries, changes over time. The challenge is not limited to understanding the strategic landscape of positioning against competitors, but also observing and interpreting the opportunity landscape. Large firms face this challenge, but do so within a generally well-resourced and often well-specified competitive context. Small, growing firms generally lack resources, especially time and industry-specific experience. And they may face these new challenges uninformed by historical context or reflected knowledge from competitor actions. The dynamics of change in opportunity landscapes are especially uncertain and dangerous for entrepreneurial firms scoping novel markets or promoting untested innovations.

Academic and consulting prescriptions for how young, innovative, growing firms should plan, adapt, and act seem to go through cycles. In good economic times when funding is widely available and technological change rapid, entrepreneurs are encouraged to think big, expand their horizons, leverage capabilities, and challenge industry norms. When economic indicators drop, entrepreneurs are exhorted to run lean, stick to their knitting, and focus on cash and low-hanging fruit. Is entrepreneurial success really just a function of exogenous context?

It may, and should seem strange that entrepreneurs should formulate their core principles around external environmental conditions. After all, entrepreneurial and innovation theory emphasizes that entrepreneurs are agents of change. Innovative entrepreneurs just think differently. In the context of information overload, which shows every sign of increasing, entrepreneurs apply counterintuitive and even counterfactual heuristics.[5]

It is the narrative framework that brings these seemingly contradictory elements back into alignment. In fact, the discrepancies between "lean" and "vision-driven" entrepreneurship are resolved through the insight of inspiring narrative. The details of the prescriptions may seem dramatically different, but there is a higher-level heuristic at work. Once again, it is the heuristic of coherence. With this tool, we can clarify how entrepreneurs manage change, address nascent opportunities, and achieve the unexpected in both lean and munificent contexts. They do so by harnessing the characteristics of the environment that support their narrative, and rewriting the narrative to accommodate the elements that conflict.

Related research suggests that large firms move towards coherent strategies in order to present the most efficient and effective interface with the market and thus compete from a position of strength against other firms in a given industry.[6] While some aspects of this process bear similarities to entrepreneurial narrative, three important differences should be clarified. First, large organizations are generally and primarily focused on optimizing the organization's exploitation of a specified and often fixed opportunity. Second, the processes associated with this type of change towards fitness are primarily reconfiguration, adjustment, adding, and culling specific resources and capabilities in the context of a working system. Finally, they rarely address reinterpretation of relationships or configurations. These distinctions are important, because the processes employed by innovative entrepreneurs to bridge opportunities are not generally available to larger organizations targeting fixed opportunities via strategic initiatives. While future research could reveal where and how larger organizations differ, for the moment it seems to be the difference between implementing strategy and narrating a business model.

Living the story

Research on venturing and entrepreneurship has focused on resource acquisition for very good reasons. Although markets are excellent facilitators of trade, entrepreneurs know that matching resources to new ideas and opportunities is anything but efficient. Entrepreneurship has even been famously characterized as the pursuit of opportunity without regard for resource requirements.[7]

The reality, of course, is much more complex. Entrepreneurs do consider resource requirements. Most entrepreneurs make careful assessments of resources available in venture planning activities. Entrepreneurs assess resource availability as a basis for adjusting the very goals they set out to achieve.[8] Scholars realized that entrepreneurs and innovators apply another tool to acquire scarce, necessary, and valuable resources. Research on passion,[9] cultural entrepreneurship,[10] and social construction[11] shows that entrepreneurs use narrative to shape how others view and assess specific venturing activities, thus legitimating the firm's goals and actions both internally and externally.

Our studies have reinforced the relevance of these storytelling and sense-making processes, but also revealed that entrepreneurs apply these capabilities beyond resource acquisition. Entrepreneurs like Matt Chambers, Pramod Chaudhari, Craig Newmark, and Matt Blumberg use narrative processes to help their firms build and cross opportunity bridges. Crossing opportunity bridges, like the initial venturing effort itself, is an example of entrepreneurial story enactment. Enacting entrepreneurial stories can be described as a three-stage process. These stages may also be linked to the coherence model associated with Insight 3.

In the first stage, writing the story, entrepreneurs generate narratives of change that describe how an innovation or opportunity will be exploited. This is, in many cases, an almost purely creative process that generates something from nothing. It is exemplified in thousands or millions of venturing events each year, big and small. Jerry Yang had the realization that, at some point, his prototype system for cataloging websites would be valuable. Writing the narrative explaining that value required an appreciation for the potential of the worldwide web and the extension of his own highly specific interests to a broader context of information access. This would have been pretty close to fiction, given that in 1994 the web was mostly a collection of strange and eclectic personal websites. After all, the original name for the website, before it was branded "Yahoo!" was "Jerry and Dave's Guide to the World Wide Web."

Dr. Brijmohan Munjal envisioned a more mobile Indian society. He created an organization that became the largest bicycle manufacturer in the world and evolved into Hero MotoCorp, the world's largest motorcycle manufacturer. At Confederate, Matt Chambers described a motorcycle company that would combine the prestige and style of high-end automobiles with the image of ruggedness and independence normally associated with brands like Harley-Davidson. At SustainableSpaces, Matt Golden created a narrative combining ecological responsibility with job creation and improving residential living quality. In this first stage, entrepreneurs imagine and tell a narrative that helps acquire key resources.

In the second stage, telling the story, entrepreneurs inspire coherence among the organizational elements and characteristics that comprise how the firm operates. Dr. Pramod Chaudhari, already a successful entrepreneur developing and constructing fuel processing facilities, imagined a story in which the need for non-petroleum-based fuels would place premiums on locally produced commodity feedstocks and the development of specialized enzymes required to efficiently process those feedstocks. His vision is being executed at Praj Matrix, an offshoot of Praj Industries, which leverages capabilities in a variety of fields ranging from chemical engineering and molecular biology to algal sciences. Throughout this and prior growth processes, Chaudhari emphasized key organizational characteristics that bind together the disparate activities, projects, and capabilities of the firm. At Return Path, Matt Blumberg ensured that through each iteration of the company's business model, key priorities and ideas remained at the core of management thinking. Blumberg and his executive team maintained the coherence of the story through acquisitions, divestitures, and the evolution of an extraordinarily complex business model that requires managing entirely distinct organizational efforts targeting entirely different customers and partners. They did so, in part, by convincing colleagues, funders, and customers that the smaller inconsistencies and incompatibilities represented relatively low-cost issues outweighed by the grand vision and mission of the organization.

Because business models are cognitive maps, this second narrative stage of storytelling is one of active sense-making. In the coherence map of the CDI business model, the critical variables are the interactions between key elements. The complementarity or conflict between variables, and the strength of those relationships, are *interpreted* and *understood* by organizational participants. Certainly, some types of interactions are objectifiable and verifiable, or subject to limited interpretation. For example, the carbon-fiber and titanium components of the Confederate Wraith are fundamentally expensive, even by high-end motorcycle standards. For someone within the organization to argue that machining key components from blocks of aluminum was either inexpensive or simple would defy imagination and might be a clear indication of cognitive dissonance. At the same time, Chambers has used these very facts to weave a narrative of unrelenting drive for perfection in the context of both performance and creative inspiration. It is a fine balance to maintain, especially when manufacturability drives a significant amount of costs beyond sourcing and machining. But Chambers' story has bound all of the compatibilities, and incompatibilities, into a compelling and powerful story.

The third stage, living the story, is where entrepreneurs achieve unexpected results. Entrepreneurs and entrepreneurial managers must do more than simply generate a convincing narrative and then organize the interpretation of priorities. Truly extraordinary, unexpected outcomes derive from entrepreneurial skills, combined with some luck, applied to bridging opportunity gaps at critical moments.

Three characteristic aspects of this fascinating phenomenon should be considered.

First, it does not appear to pre-eminently center on crisis or revelation. Although in nearly all of the case studies, founders and CEOs referenced specific periods of time in which they adjusted mental models, relatively few could be traced to a specific incident. It is important to appreciate this even in the case of CDI. The coherent model transition is motivated by the addition of the novel technology platform, but the team was well aware of the advent of the technology, and the executive level discussions about various structuring effects had been discussed previously. The final decision to merge the organizations was not specifically tied to either the university-based innovation, a sudden awareness of the potential impact of that innovation, or even the licensing of the technology from the university. Because the simulation necessitates an event-based model, it is easy to interpret this as a sort of critical incident or tipping point.

At most of the firms, as at CDI, we found that bridge-building and bridge-crossing by living the story appeared to flow seamlessly from prior activities. There is almost certainly some post-hoc rationalization involved, as individuals apply a sort of "narrative necessity" to the mental recapitulation of events. But even this helps emphasize that truly irrational and incomprehensible changes are both unlikely to transpire or succeed. Unexpected results stem from explainable circumstances, because entrepreneurs incorporate all aspects of the business model, that is, the resource, transactive, and value structures, including the people, into the new narrative direction.

Second, entrepreneurs engaged in living the story do not appear to force interpretations, conclusions, or directions on individuals or entities. Even at Savage, where the founders' long-term vision for content development appeared to be at odds with the reality of a contract shop, dozens of interviews suggested that while many employees could see the disjoint, they were under no illusions about the nature of their work. The situation was coherent – it had a plausible explanation based on the prior experience of the firm and the need to maintain multiple revenue streams. The more successful bridging companies, such as Recurve, Return Path, Voxel, and Praj, all presented opportunity spans that appeared to fit flawlessly with the prior business model.

Another example of possible interpretive dissonance was probably C5–6 Technologies. The underlying business model structures of contract research and novel product development created complex interactions between market focus and innovation focus. C5–6 built a partial bridge towards geography and feedstock specific opportunities via a commercial joint venture, but does not yet appear to have crossed that bridge.

These examples suggest that firms may be unsuccessful achieving unexpected and extraordinary outcomes when entrepreneurs are unable to facilitate crossing opportunity bridges. In addition, even when the story is coherent, it may be relatively unstable or vulnerable to becoming trapped in an energy well, as we'll discuss below. In these cases, building and crossing bridges to new opportunities may simply be impossible, because the organization has, effectively, limited its access to the necessary resources through the assumptions of its coherent narrative.

Finally, living the story appears to be an intensely personalized process for the entrepreneurs. Throughout our interviews, we found that the individuals were tightly bound to the opportunities their organizations targeted. It is interesting to note that of the case studies, the two that had the most difficulty bridging to new opportunities were Savage and C5–6. This may seem straightforward in the case of C5–6, in which an outside CEO was brought in without prior technology experience or affinity in the relevant field. But Savage provides a much subtler example to consider. Morten and Sonny left Activision specifically to start Savage based on their affinity for their work. They were clearly dedicated; numerous employees commented on the founders' extraordinary efforts and dedication as role models, their willingness to "work in the trenches" during projects, giving up salary to keep the company afloat during lean cycles. But a decade is a long time to do a job that is, in fact, slightly different than the job that was intended. One of the most fascinating comments on the Savage story came from an employee who had only been with the firm a few years, but observed:

> Sometimes I think we are fighting a noble but ultimately losing battle. It's actually what we're really good at. We'll go down fighting.

This is not to suggest that unguarded optimism is a necessary component of successful organizational narrative. Rather, the quote is indicative of the strain on the coherent business model over time. It was a strain that was evident in discussions with the founders, who talked candidly about the difference between the energizing nature of the first few years, even through extraordinary challenges, and the despiriting effect of the most recent years even as the company experienced its most rapid growth in sales, employees, and capabilities.

For most of the other companies, the link between the entrepreneurs and the opportunities was almost palpable at times. These were individuals who identified with the opportunities and who in many cases measured their lives in no small part against the realization of those opportunities. In our previous book, *Inventing Entrepreneurs*, we noted that technology affinity was a dual-edged sword. On the one hand, it ensured that the entrepreneur was enthusiastic and committed to facilitating innovation adoption. On the other hand, affinity may cloud judgment and lead to knowledge traps in which the entrepreneur is unwilling to trade the perceived failure of a given innovation for potential success in another area. We saw a minor example of this at one of our study firms. A scientist with an enabling technology he had developed himself was unable to come to terms with an alternative, more economical enabling technology that was adopted en route to full commercialization.

The cautionary note on technology affinity remains relevant, but is tempered in the context of achieving the extraordinary. The key differences between the study examples in our prior versus current research are the capabilities of the entrepreneurs and the stage of opportunity enactment. For most academic entrepreneurs, the greater danger lies in overconfidence in a technological innovation that, in most cases, was developed via the pursuit of knowledge and solutions without a market focus. In contrast to those inventing entrepreneurs, nearly all of the case studies described here are characterized by market- and solution-centric thinking. In other words, achieving the unexpected doesn't inherently derive from discontinuous or radical innovation. It results, in part, from novel thinking about how innovations are linked to markets.

Matt Chambers wanted to promote aircraft quality materials for every component of every Confederate motorcycle, but product and organizational narrative required a broader vision of rebellion as an art form. Further proof comes from the company's recent efforts to incorporate some design for manufacturing processes. Chambers describes emulating certain aspects of Porsche's production model, which balances craftsmanship with efficiency. SustainableSpaces had unique capabilities that could dramatically affect residential energy efficiency, but founder Matt Golden quickly learned that the auditing and construction skills simply weren't scalable. The company, as Recurve, is effectively selling those very same skills to competitive firms to expand its footprint and global impact. Golden loved the original opportunity so much he was willing to let it go.

Storytelling loops

Entrepreneurs who create and enact powerful narratives within organizations are interpreting and forming the firm's interpretation of the extant business model. A business model has sometimes been described as the narrative story of the business itself.[12] There are, undoubtedly, narrative aspects of the business model development process, and of course entrepreneurs often find themselves explaining, if not creating, a firm's business model in the context of telling a story to potential partners or investors. But the *story* itself is not a business model. The story is the simplification or contextualization of the business model to facilitate communication of the narrative. But the process is relevant, because entrepreneurs conceptualize, test, and refine business models through storytelling.

Organizational storytelling is a process in which individuals make sense of the complex dynamics at play in any business.[13] In large organizations storytelling is primarily a retroactive assessment: mythmaking, historical re-enactment, purpose-finding, and justification. But at entrepreneurial firms, a significant amount of the storytelling has to do with what the firm has *yet to accomplish*. The other elements are clearly in play as well, as when long-tenured employees reminisce about the trials of early start-up years or reverentially justify actions of departed founders. But entrepreneurial firms seem to live on speculation, especially in capital intensive sectors where venture financing will be needed. Senior managers find themselves constantly explaining not so much what the company *has* accomplished but what the company's accomplishments can be. The "elevator pitch," in which an entrepreneur must attempt to explain a firm's long-term value in the time it takes to ride in an elevator with a busy venture capitalist is the ultimate refinement of this process.

There are good reasons for entrepreneurs to tell these stories. Innovative entrepreneurial firms seek legitimization, either by assimilating the firm within the environment or changing the interpretation of the environment to rationalize the firm's value structure. Legitimization operates at the firm level as the narration of the business model.[14] We might dismiss this as the organizational version of "the power of positive thinking." But entrepreneurship is about acquiring critical resources, and compelling stories is one currency for achieving legitimacy. Most entrepreneurial opportunities start with one person and one idea; entrepreneurs must propagate the relevant narrative to draft others.

Our studies of entrepreneurial firms support this interpretation. In particular, there seem to be three primary storytelling "loops" that entrepreneurs use to develop a business model. The first is introspective: entrepreneurs consider opportunities and then tell a story in which they create or grow an organization to exploit that opportunity. Then the entrepreneur compares the visionary result against her own expectations. The analogy here is a golfer lining up a putt, swinging through and imagining the ball rolling forward based on the careful assessment of swing power and the condition and slope of the green. This process has sometimes been referred to as "enactment," in which the entrepreneur attempts to model the world and how things will turn out.[15]

A second storytelling loop takes place within the organization. The entrepreneur has amassed resources, including employees, and communicates a narrative version of the business model as a combination of goal-setting and motivation. Common practices for this include "all-hands" meetings, various company celebrations, internal wikis and blogs, and the variety of informal communication processes across the organization. Feedback comes in a variety of

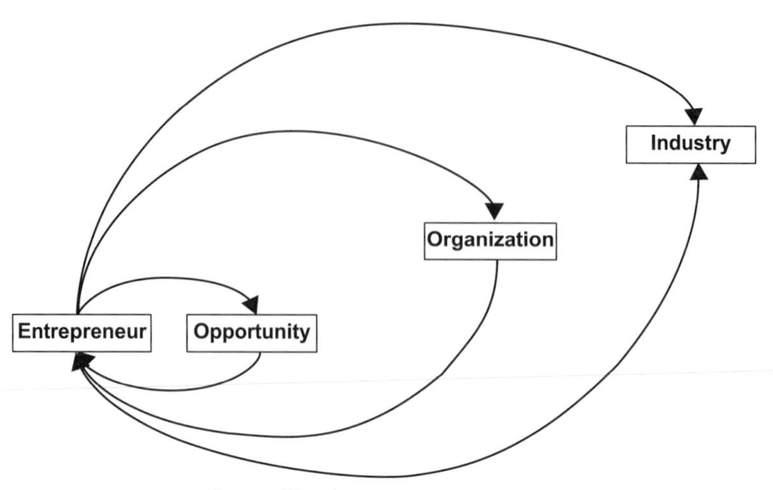

FIGURE 5.2 Storytelling loops

forms. Most of the time this can be informal, but we've seen highly formalized versions as well.

The final loop involves, of course, interaction with industry at large, whether true outsiders or allies such as partners and investors. This isn't limited to firms seeking venture capital: Savage Entertainment, the videogame developer in Los Angeles, operated for 11 years with no outside funding, but constantly told its business model story to potential and current vendors, customers, and employees. The business model story that Savage conveyed to other people and organizations evolved from a firm funding original intellectual property with contract work to a firm specializing in "better, faster, cheaper" contract work developing other companies' game ideas or porting those games from one gaming platform to another. And an important part of the evolution of the story was the simple reality that selling contract work benefited from a more focused sales story, in which the founders touted their successful prior work, rather than talking up their longer-term hopes to develop original games, especially as the company matured without bringing new games to market. Figure 5.2 shows the three concentric storytelling loops that operate in entrepreneurial contexts.

Why are storytelling loops important? Because, simply put, they don't operate independently. The idea that business models are expressed as narratives suggests that different narratives could be told to different audiences. While this is certainly true, it carries the possibility that those narratives could conflict. Narrative dissonance, as we've discussed, has the potential to detract from resource leveraging and the firm's ability to adapt and evolve. Organizations are complex entities, but surprisingly good at propagating information. And, as we've noted, humans are extraordinarily sensitive to flaws in narrative structures. When the storytelling loops aren't aligned, people will notice.

We appreciate that cases of perfectly interconnected storytelling loops are likely rare, but we've certainly seen a few. Confederate and Recurve demonstrated unusually strong coherence across all storytelling loops, but the champion in our sample was clearly Return Path. The company's ongoing war on spam and the dark side of the Internet has generated benefits for the firm, its customers, and the larger email community, but not without some casualties. On Thanksgiving Day in 2010, the executive team at Return Path discovered that the company had been hit with a phishing attack.[16] This was precisely the type of security risk that the company's new product directions target. Perhaps it isn't a surprise, however, that despite the holiday and the potentially bad PR associated with being victimized in this way, Return Path was posting information to its public blog within 24 hours. Consistent with the firm's leadership role and commitment to transparency, CEO Matt Blumberg wrote:

> [T]his phishing campaign has been going on for several months now and has claimed several victims independent of Return Path's data being stolen; however, we are acting as if any of the current phishing attempts on any of the names taken from Return Path are partially our responsibility.

There is no doubt that the firm's unorthodox approach to winning the war on spam email has been directly responsible for its success to date, but another key factor is the consistency of the storytelling loops within and without the organization. Amid the financial downturn, the phishing attack, and no shortage of other challenges, the firm has continued to grow revenues, profits, and FTEs. They are tackling entirely new challenges, such as graymail, pixel tracking, and ever more sophisticated security tools to come.

Co-authoring the narrative?

We discovered one more phenomenon associated with bridge-building and bridge-crossing. A few of these narrative transitions were unique individual efforts, such as the extended, evolving narrative developed by Chambers at Confederate, Elkins at Broadjam, or Chaudhari at Praj Industries. But we were somewhat surprised to find how many of these unexpected stories seem to stem from the work of partnerships and extended teams.

We were surprised because narrative is generally perceived to be an individual, individuating process. Most books, poems, and even songs have one author; most opportunities are ultimately identified by one individual. In addition, the challenge of developing coherence would seem to be facilitated by a unique perspective that interprets context and circumstances, configures a plausible configuration, and then conveys that narrative. This is the sort of storytelling context that has survived for thousands of years – from the oral epics of the Greeks to the wandering bards of the Middle Ages, and even the solitary sages of Hinduism.

Is there something unique and different about business narratives? Or is there something changing about society and the way in which we, as people, mediate society via business endeavors?

These are very big questions – too big for this book, in fact. We take one moment, however, to speculate. First, business has, perhaps more than any other human activity, undergone dramatic socially connected and connecting change in the past 20 years. The same factors that are driving the new entrepreneurial communities are linking those communities at individual, group, and organizational levels. While innovative ventures facilitated the application of information technology to social networking, businesses of all types are now taking advantage of the global connection. Some firms establish a Facebook page before or even in lieu of a website.

Second, the complexity and amount of available information has grown far faster than any one person could hope to track. Expert- and knowledge-management systems try to provide logic and heuristics for sorting this information, but it is unclear whether any are capable of replicating, at levels of sophistication millions of times more complex than our demonstration coherence model, what the human mind accomplishes almost instantly. So perhaps there are advantages in partnerships and teaming that enable more effective opportunity identification and realization.

If this is the case, then, why aren't there large entrepreneurial teams? We saw numerous partnerships: Vinnakota and Adler at The SEED Foundation; Dutt and Smith at Voxel; Blumberg, Bilbrey, and Sinclair at Return Path; Tom and Robert Palay at CDI. Even temporary partnerships, such as Golden and Mukherjee at Recurve, seem to work. But we saw no large, extended teams of entrepreneurs in these contexts. We can only speculate that while intellectual and narrative processing benefits from multiple sources, the communication of those processes and outcomes remains somewhat difficult and time-consuming. Thus the idiosyncratic processes that would characterize large group coherence-building ultimately reduce the effectiveness of the communication among the group. Perhaps there are significant limitations in the central scale of entrepreneurial activity of the extraordinary type. Just as many chefs spoil the broth – perhaps, too many authors spoil the narrative.

At Return Path, however, senior management decided to initiate a company-wide process called Return Path 4.0 to develop the company's fourth business model iteration. Every employee from the CEO to office assistants was placed in a group and assigned tasks associated with identifying where future opportunities for growth

BOX 5.1 Co-creating narrative: How 150 authors co-wrote Return Path 4.0

When Return Path set out to define the next iteration of its strategy, the company involved all employees in the process. Return Path 4.0 consisted of a dozen different teams of about a dozen randomly selected employees across all levels, jobs, and offices, each led by a mid-level manager. Each team was assigned a specific topic that related to the company's future growth strategy – things like "Going International," "Selling our Solutions into the SMB Market," or "Solving the Phishing Problem." The teams had four weeks and full latitude and budget to arrive at a crisp, ten-minute presentation as part of a fast-paced all-hands meeting where every team made the case for its topic. After the presentations were over, all of the backup data the teams collected was organized and delivered to the executive team to synthesize. A couple months later, the executives presented a tight, prioritized blueprint for the company's future at the year-end annual business meeting. According to CEO Matt Blumberg, "The company was electrified, engaged, and completely bought into the new strategy since every single person had contributed to its development and seen it evolve as the process went along."

would be and how the firm would address them. See Box 5.1 for more details on how they did it.

COHERENT NARRATIVES

The three stages of narrative action: writing the story, telling the story, and living the story, all function in the context of coherence forming and adapting processes at innovative entrepreneurial firms.

When entrepreneurs build narrative bridges to new opportunities, they use two coherence-seeking processes within and across those three stages, especially in the later stages. First, entrepreneurs identify and minimize frustrations within business models. Second, they reprogram the goals and priorities that determine business model relationships.

Identifying and minimizing frustrations

Bridge-building balances stability and change. Entrepreneurs reconfigure and redirect organizational elements in the pursuit of novel and possibly unfamiliar goals. This may involve minor or significant changes in business model structures already in place at the company or developing entirely new structural elements. This process is neither straightforward nor simple, in part because of unexpected outcomes. Systems theory shows that small changes in complex systems often have non-obvious consequences, some of which may also be irreversible.

As the entrepreneur redesigns the firm's business model, her first priority may be to identify and minimize frustrations between elements. Insights 2 and 3 showed that most systems are inherently frustrated. Perfect alignment among all business model elements is rare or impossible for any reasonably interesting, innovative, or complicated organization. It is often impossible to configure systems to always inhibit relationships between conflicting elements and reinforce relationships between complementary elements.

The strategic complementarity framework requires eliminating imperfect relationships. The manager patches or excises problematic elements inside the organization and changes the firm's strategic position to maximize fitness with the environment. But entrepreneurship can't be approached with the same heuristic. Entrepreneurs and entrepreneurial managers must approach coherence within a framework of plausibility rather than perfection. After all, perfection, as fitness, may be unknowable in new or nascent industrial contexts. In addition, the resources required for reconfiguration are often extremely scarce or simply unavailable to the entrepreneur.

Skilled entrepreneurs identify and then minimize frustrations within organizational systems, sometimes even using those frustrations to further broader goals. Let's take a quick look at a couple of examples.

First, let's look at how Recurve and Metalysis manage configurations that include inherently conflicting elements. Recurve is transitioning from an energy efficiency and construction remodeling firm to a software company. In doing so, however, it retained the auditing and constructing operations to serve as a combination of R&D laboratory and market-facing test site. At the same time, the auditing and remodeling business is expected to operate efficiently and, in fact, at least break even or make a profit. This defies the logic of numerous major strategic theory frameworks, including strategic complementarity, traditional strategy-structure paradigms, and transaction cost economics.

And, to be sure, there are strains clearly present and acknowledged within the organization. The technicians and contractors who work within the auditing and remodeling business are well aware that the software engineers earn significantly higher salaries. The heavy focus on routinizing the auditing process and remodeling specification algorithms to enable the software has focused the founders' efforts towards the software development process. In other words, while the auditing business remains important, senior management has prioritized software design, engineering, and testing. The attention of senior management at any organization is a precious commodity, and the auditing business simply isn't receiving as much of it.

The conflicting elements between the two operations are dramatic. The auditing and construction business combines traditional marketing and sales efforts with a highly specialized, high-touch business to consumer interface. The margins are modest and require careful, constant project management. Coordinating resources is a 24/7 job, as described by the general manager who told of an allocation process that was fluid right until the time that the trucks are sent out to specific worksites. The auditors and contractors are highly specialized in residential construction and high-efficiency residential energy materials and technologies, but they receive special training to provide a smooth and trust-generating interaction with customers, as a significant amount of business comes from word-of-mouth.

The software business operates very differently. Extensive planning and design work preceded the development and programming of a completely unique software package. The body of tacit and explicit knowledge in the founders', auditors', and remodelers' minds, based on the audits of thousands of idiosyncratic homes, needed to be translated into algorithms that could then be applied effectively to entirely new home configurations. Golden and Mukherjee brought in a high-powered software team from Silicon Valley. This team operated almost entirely within the office environment, on its own schedule, at times making demands of the auditing and remodeling business for information and resources, especially time, that effectively represented opportunity costs for that group.

The link between the software and auditing/remodeling business remains a significant conflicting relationship. But Golden and Mukherjee felt it was essential, because the spillover knowledge gained by the auditing/remodeling business facilitates the successful implementation and updating of the software. In addition, the software was being tested, sometimes in real time, at auditing and remodeling sites. In fact, this was one of the critical frustration minimization mechanisms the executive team used. It sent members of the software team out with the auditing and construction crews, sometimes to actively test software functions, but often simply to record information and observe processes. In particular, these "ride-alongs" were scheduled relatively far in advance, and in many cases the software or administrative team member would, in fact, travel to the sites with the auditing team.

Golden and Mukherjee explicitly acknowledged the conflicting relationship between the major elements of the organization.

But rather than eliminate that relationship or attempt to change the underlying elements, which would have changed the underlying business models, they minimized the frustration in order to subsume it to the larger goals of the company. Another excellent example of this was the all-company morning meetings, which were less about managing the business and more about reinforcing solidarity and common goals. One source of frustration within the auditing/construction group was that by the time they departed the company's main facility for customer sites, having already been "on the job" for an hour or so, the software and administrative teams were only just arriving. So the executive team arranged for early morning meetings once a month or so that ensured that the entire organization participated. These tended towards coffee-and-donut events rather than actually accomplishing significant work, but it sent an important signal to the entire organization: the company is bigger than the frustrations within it.

At Metalysis, the conflict was not as obvious from the outside, but was arguably much more complicated. Unlike our other study companies, Metalysis was still operating at a crossroads. At the time of the study it was perfecting the energy-efficient and low-pollution FFC process to refine high-value metals such as titanium and tantalum. Many high-technology applications in aerospace and industrial and consumer electronics require low-weight or high-specification metals. The Kroll refining process, which hasn't significantly changed in more than 50 years, is capital intensive, energy intensive, and not particularly ecologically friendly.

But Metalysis was blessed, or perhaps cursed, with multiple opportunity landscape options. First, the refining process differs for each metal, as well as the grade of the metal to be produced. There were few economies of scale associated with working on more than one metal at a time. Second, the longer-term business models for titanium and tantalum, much less the more exotic metals the company was considering, differed significantly. Titanium production is measured in thousands of tonnes; tantalum in thousands of pounds. The difference in refining capacity capital investment would also differ by three or four orders of magnitude. If the company committed to one metal, imprinting and capability effects associated with scaling up or investing in manufacturing or partnering could negatively impact its development of the other. This challenge is made more significant because most high-value metals are produced at multiple specification grades, and the decision to focus on one grade or another would inhibit its ability to develop other grades, and would likely result in how the firm was perceived in markets across different metal types. If it focused on lower-grade titanium, for example, it would create higher hurdles for its efforts to market higher-grade tantulum. But if it focused on the higher grades, it could limit its ability to find a partner for high-volume titanium production.

Initially, Metalysis management had established relatively separate teams for the development of the chemical and engineering processes required for the different metals. The CEO at the time saw both that while the inherent frustrations between the underlying elements at the company couldn't be disaggregated, there had to be systems in place to minimize those frustrations and preserve the organization design. First, while the company retained the distinct specialization between the metals teams, it overlaid those developmental teams with market-facing product managers. Rather than a true matrix structure in which individuals report to two separate managers, this cross-capabilities arrangement added relationships across the development teams. Why was this helpful? First, the company had a very limited number of test cells available for its ongoing optimization research. And once an optimization experiment had been started, that cell was unavailable for any other process for a couple of days. So prioritizing research goals and processes was essential. Second, the results of each experiment were being propagated across the organization via multiple communication mechanisms, both through the metals research groups and the market-facing channels. Senior management obtained a more realistic perspective on multiple research objectives and outputs, enabling clear directives and goal-setting.

But one of the most interesting initiatives that senior management put in place was an innovative competition within the organization. The company solicited innovations across the organization, targeting methods and ideas that would benefit the company's research, development, and commercialization timeframe. Prizes, including cash, were handed out to the most interesting and significant innovations. One year, the winner was one of the lowest-level technicians, who had been originally hired purely for manual labor. Without any formal education in metallurgy or process engineering, she identified an innovation that dramatically decreased the time of the final processing step, which had been a bottleneck for being able to turn over the processing cells. Metalysis has found ways to celebrate innovation across the organization, which has minimized the inherent frustration of operating at a crossroads.

Is it working? In 2009 Metalysis raised £5.1 million in financing and was named a Global Cleantech 100 Innovator. Between 2009 and 2010, the company doubled in size. More recently, the company initiated efforts to commission a production plant for tantalum and design a production plant for high-value titanium products. Long-term success is not assured, but Metalysis' hard work at the crossroads appears to have unexpected results: it is moving forward on multiple product types, and may be able to bridge to multiple opportunities simultaneously.

Metalysis and Recurve demonstrate that entrepreneurial firms can embody multiple, sometimes even conflicting business models at one time, if managers find ways to minimize rather than eliminate frustrations.

Reprogamming goals

Sometimes, however, firms are attempting to accomplish something so radical that minimizing frustrations simply won't suffice to create a coherent story. In our study, three firms used the more powerful tool of reprogramming goals to achieve the unexpected.

Confederate CEO Matt Chambers reformed the company out of bankruptcy and rewrote the business narrative. The company would be design-led and philosophically centered. Traditional business functions would be secondary to these purposes. This is, in many ways, an extraordinary accomplishment, because Chambers has managed to keep multiple, seemingly paradoxical priorities and heuristics in play.

First, Chambers operates a for-profit company that has, in fact, utilized a variety of financing mechanisms, all of which place various financial obligations on the organization. Second, Chambers himself argues passionately for a philosophy of objective humanism, which combines the ultimate explanatory power of human reason with the individual rights and laissez-faire economic policies most commonly associated with the author, Ayn Rand. And he promotes Confederate as an organization on a mission to promote the expression of the strength of American industrialism and rebellion as an art form. He has successfully reprogrammed the goals of the organization, and the employees, customers, and stakeholders of the company, to share these priorities regardless of whether Confederate ultimately succeeds as a for-profit firm. In one interview, a Confederate employee talked about eventually being able to purchase a used or damaged Confederate bike and rebuild it. With starting prices ranging from $45,000 to over $100,000, the company's products are out of reach for most of its employees – but they have bought into the passion that sustains this company through recession, bankruptcy, and hurricanes.

At CDI, the stem cell company, the goal reprogramming has been much more subtle, but equally important. Merging the distinct organizations required difficult decisions and the creation of new frustrations among organizational elements. The CDI executive team established an organization-wide innovation project, but specific to identifying novel market targets based on the new technology uptake. In doing this, the company explicitly clarified that preliminary successes in scaling up the iPS technology meant that the company would move forward on a single platform, even though that meant driving different business models: research tools for drug discovery in the short term and therapeutic products in the long term. It is interesting to note that whereas Metalysis became more market focused in its explicit internal philosophy, CDI became more technology focused. This is non-obvious, because CDI has been actively ramping up its market-facing structures, including the development of its sales and marketing team, while consolidating its R&D activities and focusing manufacturing on scale-up. But the intermediate goals have been centralized around the new technology platform, enabled by unanticipated advances in scalable production processes. The firm remodeled its coherent story, despite significant internal conflicts, and arguably entirely distinct business plans welded together, around the iPS technology platform.

A fascinating example of goal reprogramming is presented by well-known internet firm Craigslist. The growth and success of the organization has occurred even as, and perhaps because founder Craig Newmark refused to fully monetize the network potential enabled by one of the most visited and connected internet sites. Despite being the largest source of classified ads in the world, Craigslist only charges fees for a very small subset of its services: job placement ads in select cities and apartment listings in New York. Craigslist CEO Buckmaster stated that Craigslist has no plans to maximize profits, as became clear in a meeting with Wall Street analysts:

> Wendy Davis of MediaPost describes the presentation as a "a culture clash of near-epic proportions." She recounts how UBS analyst Ben Schachter wanted to know how Craigslist plans to maximize revenue. It doesn't, Mr. Buckmaster replied (perhaps wondering how Mr. Schachter could possibly not already know

> this). "That definitely is not part of the equation," he said, according to MediaPost. "It's not part of the goal."[17]

Craigslist is now being challenged by more traditional internet monetization model businesses such as Oodle and Kijiji, and only time will tell if it is forced to give up its non-traditional philosophy and goals. But it is hard to argue with the success of the company to date. Reprogramming organizational goals can create powerful narratives within organizations that may defy even the most broadly based conventional wisdom.

BRIDGING NARRATIVES

Most opportunities dissipate over time – some ever so slowly, others very rapidly. As the discussion on Insight 4 highlights, large firms bring to bear time, money, and expertise to innovate business models by simplifying organization design and increasing attention to opportunity landscapes. But smaller firms rarely have the luxury of such assets. When small entrepreneurial firms span opportunity gaps, they often do it by leaps that seem fantastic or impossible. And it is the power of narrative that helps them do it. Even so, our investigation revealed the activities that talented entrepreneurs initiate during these transitions.

> First, entrepreneurs rewrite the organizational story.
> Second, they redraw the organizational system.
> Third, they inspire the narrative to bring along key stakeholders.

Rewrite the story

We've already seen numerous examples in which entrepreneurs use narrative in their efforts to change organizational processes and activities. But when entrepreneurial firms must undergo dramatic transitions, the narrative entrepreneurs use is often rewritten from scratch. In the early years at Broadjam, Roy Elkins imagined that the firm would be the nexus of music distribution by consolidating and then marketing unsigned musicians' work via the big peer-to-peer networks like mp3.com and Napster. But ultimately the legal institution caught up with the new technology, and Broadjam's entire distribution model, and thus its marketing power to independent musicians, was effectively eradicated.

Elkins looks back and refers to that period as both a "dark time" for the organization but also a revelation for himself and his team. It revealed the nature of the underlying goals that Elkins had in mind – helping independent musicians monetize their skills. There was a time when Elkins believed that Broadjam could compete with the recording studios. But the elimination of the fully disaggregated music distribution system, which was predicated on unfettered access to music without a clear pricing policy, clarified for Elkins that his goal, and Broadjam's purpose, was more about enabling hobbyist musicians than reconfiguring the music industry. Well in advance of the iTunes launch, Elkins rewrote the Broadjam story to more explicitly center on the independent musician. In the process, Elkins reinterpreted every aspect of the firm's activities and projects to fit that long-term goal. The consulting projects enabled and informed more effective cataloging and metatag mechanisms. The online competitions leveraged tools that would eventually help hobbyists promote and get feedback on their music.

An interesting aspect of the rewrite emerged during our study period – as the Broadjam team turned its attention to the artist fan base. Just as the organization centered its music distribution around artists, rather than the commercial sector, Elkins knew that for many of its artist clients, a fan base was as rewarding, both financially and psychologically, as the commercial contracts the business facilitated. A visit to the new Broadjam website reflects the continuing evolution of the story, as nearly every feature and structure centers on the artists themselves. The commercial opportunities are easily accessible: licensing, contests, and downloads. But it all comes back to the artists. The system directs users to the artist, and enables a social environment centered on each artist and that artist's music. That focus can be traced back to when Elkins rewrote the entire Broadjam story at a time when it wasn't clear the company would survive.

Broadjam was one of the dot-com music companies founded between 1997 and 2000 that hoped to cash in on the temporarily broken musical access infrastructure. But the narrative bridge that Elkins built is the difference between being one of dozens, perhaps hundreds of firms that have subsequently failed, and being one of the few, and possibly the only one that survived.

Redraw the system

Bridging narratives requires narrative coherence, but it is a more dramatic change than coherence-seeking. In these cases, when organizations adapt to new opportunities, or create new opportunities, entrepreneurs redraw the entire organizational system.

For clarity, this may include significant changes in the formal structure of the organization, including the "organizational chart" of formal hierarchical relationships, but it may not. Redrawing the system is about rearranging the key elements in the firm's business model to change the fundamental interpretation of the firm's resources, processes, and goals.

Voxel redrew its business model in transitioning from a server capacity business to a content delivery network. Rather than emphasize the hardware infrastructure, the firm needed to refocus on an integrated solutions mentality that drew heavily on the real-time capabilities of its software and network engineers. To do that required putting those engineers to work on the front line, interacting with customers, rather than in a back room development and implementation model. This was, in many ways, a terrific risk: it challenged the capabilities of the individuals, put significant strain on the company's administrative systems, and gambled on the abilities of the engineers to more cost-effectively assess problems and design solutions than would be possible via a traditional customer service organization.

It worked. And while there are questions on whether this can be scalable, something being tested in the market as we speak, it helped Voxel achieve the unexpected – the fastest content delivery network in the world, with only a fraction of the resources and infrastructure.

Inspire the narrative

To some extent, all of the stories we've retold about entrepreneurs achieving the unexpected have relied on the ability of those entrepreneurs to inspire narratives within their organizations. In extreme cases, we observed almost cult-like reverence for the founders and CEOs. One employee described the highlight of his year being the day when the CEO invited him to chat, just to check in, and remembered the names of his family members.

But it would be wrong to suggest that traditional characteristics of charisma provide the foundation for narrative inspiration alone. To be certain, many of the entrepreneurs in these stories are charismatic. And they have had years, in some cases decades, to hone their leadership skills. At the same time, inspiring the narrative, the process of communicating the story to stakeholders and bringing them along for the ride, is facilitated both by the storytelling capabilities of the entrepreneur and the coherence of the story itself.

Walter Fisher's research on narrative rationality suggested that the power of storytelling lay in two underlying factors: narrative coherence and narrative fidelity.[18] Narrative fidelity is the strength and reliability of the storyteller. It is the charisma, whether explicit in the media presence of someone like Steve Jobs, or implicit in the fundamental trustworthiness of the source, like former American newscaster Walter Cronkite, often referred to as "the most trusted man in America" during his lifetime. But narrative fidelity is not sufficient; in fact Fisher argued that narrative fidelity was secondary to narrative coherence, the plausibility of the story.

There are many other theories to explain the power of the story. Some suggest that storytelling impact relies in part on surprise, ambiguity, or even discordance. But Fisher's theory is particularly compelling. It relies on the idea that the human mind processes extraordinary amounts of information, but tends to rely on simpler heuristics, triggers, and patterns in drawing conclusions. And quantitative research on decision-making, especially in judgment-requiring contexts, clearly demonstrates that this process of coherence-seeking closely matches the subtle, sometimes unconscious shifts in human perception and belief.[19] Perhaps, then, it is no coincidence that neural networks demonstrate this coherence-seeking property, and that the results of our modeling work match the human interpretations and outcomes at organizations.

In the end, charisma can only take an organization so far. Leadership is a powerful tool, a lever for inspiring action and emotion. But our study of entrepreneurial activity shows that achieving unexpected results requires the creation and promulgation of creative and powerful narratives that present plausible, if sometimes unlikely, bridges to entirely new ideas and opportunities.

While all of the entrepreneurs we interviewed and observed displayed leadership skills, most did not rely on charisma as the primary tool to lead the organization. Matt Chambers was the one possible exception. His passion and personal vision were impossible to ignore. Confederate has had an extraordinary and enduring effect on motorcycle design, very much reflecting Chambers' philosophical ideals and zeal.

In contrast, while Matt Blumberg is also enthusiastic and driven, and a convincing storyteller, he is much more a father figure than a prophet. And Thomas Palay, co-founder and President of Cellular Dynamics, combines the confidence of the law professor with the competence of the corporate executive.

In the end, inspiring the narrative benefits from charisma and traditional leadership qualities, but more importantly, it is about writing inspiring stories that co-opt stakeholders and link organizational structures and activities to the underlying purpose of the business. It is how less than 100 people at Craigslist manage the world's largest classified ad system. It is how less than 300 people at Return Path influence the deliverability of billions of emails every day. It is how less than 15 people at Broadjam have helped create the world's largest catalog of independent music. It is how everyday entrepreneurs achieve the incredible.

REBEL WITH A STORY

> Being shot out of a cannon is better than being squeezed out of a tube. That is why God made fast motorcycles, Bubba.
>
> Hunter S. Thompson[20]

The narrative of Confederate Motorcycles is explicit in everything the firm does. Its website ranges from purist engineering to near-rant, in the process quoting Hunter S. Thompson, T. S. Eliot, John F. Kennedy, and Rosseau. The company unabashedly, though with an explicitly satiric, ironic streak, describes itself and its products with the words: "humility," "genius," "toughness," "kindness," "classless," and "gentlemanly."

In telling the Confederate story, it's sometimes difficult to know where to start and where to finish. Chambers was a successful trial attorney in the 1970s who underwent his own transformation in the decision to start Confederate.

> The idea was born on April the 2nd of 1991. That was when I first made my commitment to do this, when I decided that this was definitely the direction I'm going to go. It was a big commitment ... I've been entrepreneurial my whole life. I opened a pool hall in 1971, started and ran my own law practice for 13 years ... As a kid, I'd always loved cars and motorcycles, and I fiercely

> believe ... in a space where people came to empower themselves as individuals, and that this was imbued into the design of American machinery. And I felt, and I still do, that this was slipping.

It took the company more than five years to actually produce a motorcycle, the Ghost, of which it sold a few dozen while ramping up design for the more revolutionary bike, the Hellcat.

Amid the extraordinary availability of capital to early stage firms in the late 1990s, Chambers submitted to the vision of a massive expansion plan funded with outside capital. The firm took on millions in venture funds, only to see long-term financing options, along with the global economy, collapse in 2001 as the tech bubble burst. For Confederate, manufacturing expansion plans could only be supported by cost-cutting and commensurate price-cutting. For Chambers, the reality of reporting to external financiers hit home when the Board fired him in late 2000, in part because he fought against the efforts to mass produce Hellcats and to lower manufacturing standards. The company declared bankruptcy in late 2001, and Chambers was able to buy it back in a distressed asset sale. The Hellcat had been a success in its own right, garnering rave reviews for design and performance, but the compromises in the production process during the control battle meant that a significant portion of that production run would require more service and support in the long run than Confederate bargained for. A decade later, Chambers' admonition to resist compromise via mass production rings eerily true:

> We obsolete our unique individualist genius, and in the process we obsolete ourselves. Contrived system mechanics secures our deepest fear and anxiety.[21]

By 2005, however, Chambers had the firm back on its feet. Infused with his own savings and a few choice investors, the firm had brought the Hellcat back to the planned quality level, and was ramping up to begin manufacturing the Wraith, arguably the most radical motorcycle design in recent memory. For Chambers, the Wraith was the full personification of the firm – totally idiosyncratic, built of titanium and carbon fiber without compromise, presenting a purposefully skeletal image of the American South as ravaged and defiant. Photos released to media outlets were drawing first responses of shock and incredulity within the design and motorcycle communities.

On August 28, 2005 Chambers was in the Middle East wrapping up discussions with additional investors, who, like many of the firms' private investors, were also enthusiasts and owners of the company's motorcycles. The documents were drawn up, the terms were set, and Chambers prepared for the next stage of the company's growth and expansion.

On August 29, Hurricane Katrina crashed directly into New Orleans. Like most of the country, the Confederate crew hadn't realized the destructive power of the storm until too late, and they had been unable to relocate the firm's equipment and inventory. Chambers could only get information in bits and pieces from employees and friends – New Orleans had effectively been put under martial law and most of the city had been evacuated. He knew the situation was bad, but he didn't know how bad. Over the next few days, as information trickled out of the city, Chambers grew to understand that "bad" would dramatically understate the reality of the destruction. Ultimately, it became clear that Confederate's facility had been completely destroyed when the roof collapsed under the massive pressure of the rain and wind.

But like the phoenix arising again from the ashes, Confederate was already rebuilding. Despite the certainty that Katrina had effectively wiped out the company, the Middle East investors stuck to the terms of the financing deal. Confederate was suddenly a company with no assets other than the investors' cash, the dedicated team, and a proven business model. Although nearly every operational and administrative aspect of the firm would have to be redeveloped from scratch, the frames of the first two B120 Wraiths were extracted from the facility structurally intact and nearly undamaged, a testament to the engineering and material strength of the design, and a striking metaphor for the company as a whole.

During 2005, the company successfully manufactured and sold Confederate Wraiths to an adoring and highly select market. Development of an even more radical machine, the Renovatio, originally slated for manufacturing in 2009, was halted in preference to developing the Confederate Fighter, another startling titanium design that captivated the cycling world, originally sold exclusively through Neiman-Marcus.

In late 2008 and early 2009, Chambers decided the time had come to expand the vision for Confederate. A COO was hired so Chambers could focus on sales, partnerships, and fundraising. He created more administration at the organization to better support existing customers and develop a more streamlined manufacturing process. He mandated improved documentation and oversight of nearly every aspect of the firm's operations. To ensure the company's stability through this phase, Chambers raised private equity funds through a complex transaction that actually listed Confederate on the US OTC stock exchange, where it remains today.

Throughout this tale, Chambers has told a story of human characteristics that he believes should be incorporated into every product, and every human endeavor. He has rewoven the threads of the story year in and year out, to enable the radical designs of the Wraith and Fighter and now the more minimalist design of the new Hellcat. He has never defined success solely by growth and making money.

> It's fine if we don't make huge amounts of money. Don't get me wrong – this company will make profits and it will survive. But that's not why I'm doing this, and that's never what I set out to

do ... I knew that if I stayed true to the vision that things would work out ... we're only looking to be a company that, maximum, produces 300 bikes a year. I'd like to see all sorts of other products, cars, everything, made to much higher, craftsman standards. I wouldn't want to be the guy putting a washer on a bolt all day long, with things whizzing by. And I would hate to have my grandson on my lap and have him say, "Grandad, what did you do [for a living]?" and I would describe that. I know that people did that for the sake of their grandkids, but I think that's too high of a sacrifice.

NOTES

1 "Speed demon" by Phil Patton, *ID* Magazine. June 2005, pp. 78–79 (accessed via Confederate.com website on July 1, 2010).
2 As of July 2011.
3 Matt Chambers speaking at the 2011 Gravity Free Design Conference, San Francisco, May 24–26, 2011 (speech text provided by Confederate Motorcycles).
4 Kawasaki, G. 2004. *The Art of the Start*. New York: Penguin.
5 Baron, R. 1998. Cognitive mechanisms in entrepreneurship: why and when entrepreneurs think differently than other people. *Journal of Business Venturing*, 13(4): 275–294.
6 Siggelkow, N. 2002. Evolution toward fit. *Administrative Science Quarterly*, 47(1): 125–159.
7 Stevenson, H. H. and Jarillo, J. C. 1990. A paradigm of entrepreneurship: entrepreneurial management. *Strategic Management Journal*, 11(4): 17–27.
8 Sarasvathy, S. D. 2001. Causation and effectuation: toward a theoretical shift from economic inevitability to entrepreneurial contingency. *Academy of Management Review*, 26(2): 243–263.
9 Kotha, S., Chen, P., and Yao, X. 2009. Passion and preparedness in entrepreneurs' business plan presentations: a persuasion analysis of venture capitalists' funding decisions. *Academy of Management Journal*, 52: 199–214.

10 Lounsbury, M. and Glynn, M. A. 2001. Cultural entrepreneurship: stories, legitimacy, and the acquisition of resources. *Strategic Management Journal*, 22(6/7): 545–564.

11 Downing, S. 2005. The social construction of entrepreneurship: narrative and dramatic processes in the coproduction of organizations and identities. *Entrepreneurship: Theory & Practice*, 29(2): 185–204.

12 Magretta, J. 2002. Why business models matter. *Harvard Business Review*, 80(5): 86–92.

13 Gabriel, Y. 2000. *Storytelling in Organizations: Facts, Fictions, and Fantasies*. Oxford University Press.

14 Lounsbury, M. and Glynn, M. A. 2001. Cultural entrepreneurship: stories, legitimacy, and the acquisition of resources. *Strategic Management Journal*, 22(6/7): 545–564.

15 Child, J. 1997. Strategic choice in the analysis of action, structure, organizations and environment: retrospect and prospect. *Organization Studies*, 18(1): 43–76.

16 www.returnpath.net/blog/intheknow/2010/11/security-alert-phishing-attack-update/ (accessed June 1, 2011).

17 "Craigslist meets the capitalists" by Dealbook, *New York Times*. www.dealbook.nytimes.com/2006/12/08/craigslist-meets-the-capitalists/ (accessed January 24, 2011).

18 Fisher, W. R. 1987. *Human Communication as Narration: Toward a Philosophy of Reason, Value, and Action.* Columbia: University of South Carolina Press.

19 Simon, D. 2004. A third view of the black box: cognitive coherence in legal decision making. *The University of Chicago Law Review*, 71(2): 511–586.

20 Thompson, H. S. 2003. *Kingdom of Fear*. New York: Simon and Schuster.

21 Matt Chambers speaking at the 2011 Gravity Free Design Conference, San Francisco, May 24–26, 2011 (speech text provided by Confederate Motorcycles).

6 Embrace the unexpected opportunity

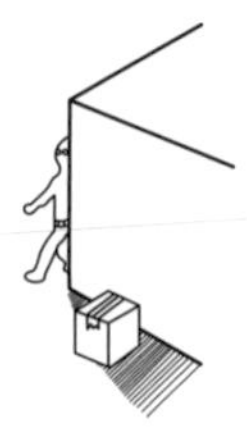

SURPRISE! OPPORTUNITY.

Where do unexpected opportunities come from?

The first hour following injury or acute disease onset is often referred to by medical and military personnel as "The Golden Hour," because effective treatment during that time dramatically decreases adverse outcomes and fatality rates. But India's vast and varied geography, high density cities, and limited transportation and health care infrastructure leads to millions of unnecessarily traumatic outcomes from disease and injury. How could government or private care effectively address this complex structural problem embedded in unchangeable factors? With the guidance and philanthropy of Mr. Ramalinga Raju, Venkat Changavalli founded the Emergency Management and Research Institute (EMRI) in Hyderabad, India in 2005. EMRI was the first emergency response provider in India. EMRI has since grown into a public-private partnership with a vision of responding to 30 million emergencies and saving one million lives *every* year.

Despite extensive awareness campaigns about hygiene in hospitals, MRSA infections continue to cause thousands of deaths in hospitals in the UK and worldwide. These are almost all preventable, because MRSA is unrelated to the hospital patient intake. MRSA is caused by bacteria resistant to standard antibiotics; infections are linked to poor hand-washing routines. A design student at the Royal College of Art's Human Design program, Adam Sutcliffe, realized an unexpected insight while touring an NHS facility. The very factors that inhibit effective hand hygiene could be harnessed towards increased hand-sanitizing behavior. He developed a small single hand-action device to deliver alcohol gel and a pleasurable tactile sensation. In a small-scale trial, nurses were observed using the device every 3–5 minutes, unconsciously replicating World Health Organization recommendations precisely.

The internationalization of communities around the world has resulted in heterogeneous schools and workplaces where multiple languages are spoken at various fluency levels. The challenges associated with communication, learning, and teamwork in these contexts have led to efforts to homogenize language instruction and encourage single-language environments. At the University of Edinburgh, Professor Antonella Sorace thought about her own research on brain flexibility and realized that a strategy encouraging bilingualism could help schools and organizations make smarter students and workers. Still hosted by the University of Edinburgh, her entrepreneurial efforts are expanding Bilingualism Matters from Scotland to Europe.

In these cases, and so many more, entrepreneurs from a variety of backgrounds see something that is unexpected and embrace it. Unexpected opportunities present varying degrees of complexity. Opportunities need not be rocket science to be surprising; entrepreneurs don't need to be rocket scientists to identify and understand them. It is about seeing something as the source of a problem, and solving it with a different approach that has not necessarily been tried-and-tested. It is about accepting uncertainty, at a level of personal risk that the entrepreneur finds appropriate, not knowing how the end product will turn out.

BEYOND UNKNOWN UNKNOWNS

Entrepreneurs face the unknown, strive to create new value, and take risks. But entrepreneurship is not all about risk-taking. Starting new ventures and reaching for the unexpected requires accepting the possibility of failure, the risk that desired outcomes won't be attained. One of the central challenges for entrepreneurs is managing uncertainty, especially when outcomes are unclear, unknown, or unknowable.[1, 2]

It is a common, intuitive myth that entrepreneurs are defined by a higher tolerance for risk than the general population. Studies have shown, however, that entrepreneurs don't like or tolerate risks any differently than non-entrepreneurs.[3] What defines entrepreneurs is not their tolerance for risk, but their approach to managing uncertainty.[4] Research comparing bankers to entrepreneurs highlights this distinction. Bankers focus on target outcomes and seek to control risk exposure associated with those outcomes. Entrepreneurs, on the other hand, choose a comfortable risk level and then focus on achieving outcomes at that level of risk. In other words, entrepreneurs see risk as a factor inherent in new value creation, and stretch what can be accomplished in that context.

Entrepreneurs also frame challenges based on their own values and assume greater levels of personal responsibility for outcomes.[5] Individuals with entrepreneurial characteristics optimize decision comprehensiveness to take on new opportunities specifically when they believe they have the capacity to manage uncertainty.[6] This is especially relevant in uncertain, turbulent, or dynamic environments.[7]

Why does this matter? It is the core of embracing the unexpected. The most innovative entrepreneurs pursue opportunities when outcomes cannot be fully visualized. They do it by managing uncertainty throughout the entrepreneurial journey. When risks are constant or stable, these entrepreneurs focus on building coherent organizations. When uncertainty is ubiquitous, they inspire new narratives that make sense of the uncertainty and lead the organization to entirely new, coherent capabilities.

Decision analytics has suggested that information, including information about opportunities, can be described as known or unknown. The unknown may be divided into "known unknowns" and "unknown unknowns." The known unknowns are factors or issues we are aware of but can't quantify or analyze. Unknown unknowns are the factors and issues that we can't analyze, at least in part because we aren't aware we don't know about them. But beyond unknown unknowns is the unknowable. These are the predictive factors that drive extraordinary innovations and unexpected outcomes.

One of the problems with studying entrepreneurs, as we have noted, is that we tend to study success stories. And, because we are narrative creatures, we find stories that explain those successes that make sense. But many of the truly incredible entrepreneurial stories of the recent past represent situations in which innovators addressed opportunities that were partly or entirely unknowable. We may think that we effectively explain them in hindsight, but that form of analysis is easy to generate and impossible to prove. Imagine, for example, if a company like Microsoft or IBM had launched the Google search engine. What if it were an Australian company? Israeli? British? What if it had been the US Government? Doctrines of inevitability are inherently false in an organizational context. There is no reason to believe that a change in any of the factors associated with the launch of Google, whether founders, location, funders, or growth processes, would have still led back to the company we know as Google.

The concept of the learning organization[8] has been a major contribution to strategic theory. Firms and even industries can and do learn. And, in part, that learning depends on what the company already knows – not just in terms of building on past knowledge but its absorptive capacity, its ability to learn similar or entirely new information.[9] To be sure, the success of many an entrepreneurial venture depends, in part, on its ability to learn, perhaps very rapidly, especially during international growth.[10] But learning curves alone can't explain these outcomes when firms have no prior learning base to build on, and when path outcomes are simply too complex and uncertain to be modeled by any reasonable means. In these situations, entrepreneurs accept that implausibility is simply a perceived limitation associated with the lack of information. They accept that by the time the necessary information becomes knowable and accessible, the window of opportunity may already have closed.

Our final observation is *Insight 6: Innovative entrepreneurs embrace unexpected opportunities.*

Venkat Changavalli left a plush corporate job as CEO of a pharmaceutical business. His goal was to organize efficient emergency response in a country with impossible traffic, and geographically and culturally diverse populations. Though much better than before, anyone driving on Indian roads would concede that it is a miracle that traffic moves along and work gets done every day. Changavalli needed to design, organize, and implement an opportunity that would have been considered implausible. He implemented the idea by starting from scratch, picking an emergency phone number 108, designing the ambulance to be low cost and suitable for Indian emergency needs, creating a hospital network, and a call center that promises that a call will be answered before the second ring. EMRI delivers this service free of cost to the customer. By 2011, EMRI operated in 11 states and attended 8000+ emergency calls per day, reaching 73 percent of callers within 15 minutes.

Adam Sutcliffe, a former web designer and brand consultant, had reinvented himself by going back to his passion for design. Completing a late Masters in Industrial Design at the RCA, he visited a hospital to examine human factors in health provisioning. He noticed that the ubiquitous presence of signs and wall-mounted sanitizers were clearly failing to change ingrained habits. Doctors and nurses were simply too rushed, too focused on minute-to-minute care issues. Sutcliffe imagined his own childhood, pondered the innate response to dirty hands, and saw the opportunity to completely change the drivers for desired behavior. Rather than require health practitioners to learn new routines, he would reward and reinforce the apparently problematic routines, subsuming them to the needed purpose. Just as children do, doctors and nurses instinctively dry wet or dirty hands on trousers or hospital gowns. So the Orbel sanitizer would be located in that position, release the antibacterial gel in response to the wiping motion, and provide a pleasurable tactile sensation, rewarding the behavior to make it habitual. Now the doctors and nurses sanitize their hands via the ingrained response learned as children and get habituated to the pleasurable sensation associated with the action. Sutcliffe's venture, Orbel Health, hopes to bring this innovation that harnesses a "negative habit" to hospital hygiene wards to fight MRSA and other superbugs.

Professor Sorace began to appreciate the potentially positive impact of bilingualism in response to the gap between her research results and the comments of parents and teachers regarding her son's bilingual capabilities. Although schools and workplaces tend to see non-native speakers as handicapped in the learning and work environments, Sorace knew that bilingual individuals demonstrate more flexible and absorptive brain capacity.

All of these are unexpected opportunities for the *individuals* who undertook them. Some or all the business model elements associated with these ventures already existed. These individuals found new combinations, new configurations, new applications, and sometimes entirely new meanings for those business model components. These entrepreneurial inspirations go beyond alertness, beyond goal adjusting associated with effectuation, beyond the extension of prior experience and capabilities. In the context of the unknowable, these entrepreneurs identified imperfect opportunities, refused to accept goal limitations, and saw through or beyond their own base. They rewrote the narrative of the opportunity by incorporating their own participation, developing stories in which they were both authors and protagonists. The most unexpected businesses have been created by entrepreneurs who brought business sense, design sense, and a willingness to believe in the power of opportunity. They embraced a strange uncertain certainty. Each accepted and cultivated the unyielding belief that despite being uncertain about the destination as well as the path, engaging with the opportunity was an essential component in its realization.

INSIGHT 6: EMBRACE THE UNEXPECTED OPPORTUNITY

Based on the importance of design, coherence, and narrative inspiration, we provide five actions to help entrepreneurs embrace unexpected opportunities.

1 *Manage dissonance*

In the context of building a coherent organization, dissonance plays an equivocal role. On the one hand, it may represent an indicator of fundamental disconnect between narrative and the business model. This was the case at Savage Entertainment. On the other hand, tension between business model elements, exemplified by the frustrated systems at CDI and Recurve, are simply symptoms of coherent configurations necessitated by opportunity unknowability. Both individually and via organizing activities, entrepreneurs make sense of the world as a critical step between information gathering, analysis, and decision-making.[11] This process is especially powerful in the context of dissonance, the conflict between apparently contradictory interpretations of the world.

Managing dissonance is the first way that entrepreneurs embrace the unexpected. We've discussed dissonance in the context of building coherent organizations. The most innovative entrepreneurs reach for the unexpected by harnessing dissonance. These entrepreneurs use the tension within the organization to retain flexibility within configurations and prevent lock-in of non-essential narratives or business model structures.

Retaining tension at Recurve has helped the company adapt to constantly changing regulations, economic conditions, and software development requirements. Without the tensions between the auditing/construction side of the business and the software development efforts, more traditional software design processes would likely have dominated, limiting the essential value and complexity associated with the idiosyncrasies of residential energy efficiency profiles. The strength of the Recurve software is that it applies sophisticated heuristics to direct assessment of individually unique residential configurations. The software will never be used precisely the same way twice, a design framework that would have stymied traditional software engineering processes. The essential role of the auditing side of the organization provided a constant tension that prevented the software engineers from settling on standardizations that would have made the software more efficient, but ultimately less effective.

2 *Unframe opportunities*

Traditional strategic decision-making is a quest to "frame" problems. Managers assess context and conditions in order to place unfamiliar situations into a familiar cognitive framework. This is, in essence, the nature of the business case study – the application of established heuristics to novel situations. More aggressive, entrepreneurial approaches have developed the power of reframing information and decisions. In this mode, the intent is to consider the relevant information, apply a framing methodology that seems relevant, and then reframe the entire problem into another context or methodology to test assumptions and outcomes.

But achieving the unexpected requires a more radical approach.

Entrepreneurs who grasp at imperfect or unformed opportunities need to *unframe* their thinking altogether. The process of framing and reframing analysis is central to most of our rational decision-making activities precisely because we have no well-developed mechanisms for decision-making under extremely high levels of uncertainty, in particular when we can't even assess how much or what type of uncertainty is in play. It is in these contexts that entrepreneurs rely on intuition as an a-rational, rather than irrational or non-rational, heuristic.[12]

Unframing the problem accepts that prior methodological perspectives or rules of business analysis may be impossible or inappropriate in the context of new market dynamics or entirely new industry systems. We might like to apply traditional organizational theories to all situations, but the successful commercialization of many recent innovations has required generating entirely new theoretical frameworks for entrepreneurial success.[13] In time we may find more generalized rules that can be retroactively fitted to both traditional strategic frameworks as well as highly innovative entrepreneurial contexts. But in a world in which the success of innovative business models simply can't be predicted a priori, the entrepreneurs who choose to enact opportunities without subjecting the challenges to the boundaries of traditional frameworks will achieve the unexpected.

Consider Confederate Motorcycles. Founder and CEO Chambers understood the characteristics of rebellion that other motorcycle manufacturers had made attractive to mass markets. But rather than compete with them, he wanted to outdo them. He wanted to rebel against what he saw as the mass commercialization of rebellion. Rather than make the identity of rebellion accessible, he would place it out of reach of all but the most wealthy enthusiasts. In effect, Confederate created a story in which radical perspective links popular culture to Chamber's economic philosophy. The meaning imbued into the organization would be hard to rationalize in almost any other context. Epitomizing rebellion with an $85,000 price tag would draw heavy criticism at a place like Harley-Davidson, with a century of motorcycling history and roots in fiscally and culturally conservative Milwaukee, Wisconsin.

But this is precisely where the idiosyncrasies of innovative organizations defy traditional strategic theory. Chambers actively reads books on strategy and entrepreneurship and applies what he thinks is relevant. But he's equally pleased to put them aside when he's convinced the received wisdom can't help him do what he wants to do. He's learned that, by definition, the right path for his unusual organization simply can't always be found in those books. The purpose, process, and values at Confederate are unique precisely because they are difficult to codify and structure into a format that can be replicated. The success of these unique firms stems, in part, from the fact that they do what hasn't been attempted, or attempted successfully, by anyone else.

3 *Create routines to generate novelty*

At its very basic definition, a routine is a process. It can be reliably repeated, and produces the desired outcome every time. It is just "routine." Now, as venturing becomes more routine, entrepreneurs benefit from proven mechanisms to build efficient and effective organizations. For example, specialized routines allow businesses to outsource sophisticated intellectual property patent prosecution to extraordinarily bright and competent attorneys, even when those attorneys have relatively limited perception into the implementation of the ideas behind the patent. That is, many of the key processes in starting and getting a new venture established have become routine! In this context, one might expect that entrepreneurs will tend towards programmed activities and focus on incremental improvements or marginal extensions of successful businesses into adjacent markets.

In fact, the opposite appears to be true. There is something extraordinary about the nature of entrepreneurship that many of the individuals who pursue it choose "the road less traveled." In interviews, we heard and saw stories of entrepreneurs experimenting for the challenge and reward of accomplishing the previously impossible or unknown. The routine stuff allows them to expand into novel areas. In some cases these entrepreneurs developed entirely new ways of routinizing, but sometimes recombined extant routines to create novel solutions.

Bharti Airtel implemented a business model of inclusive growth that brought technology, resources, and wealth to the communities it serves – the same communities that had been written off as inaccessible by multinational mobile carriers. It captured rural markets by harnessing low-cost advertising, decentralizing the distribution of mobile top-ups to the local corner shops, and focusing on customer service. Shifting the investment in costly infrastructure by leasing towers from network infrastructure providers allowed the firm to scale up faster than competitors who owned and operated their cell towers. The system worked because the routines were simple and allowed faster scale-up. The routines themselves were not new, but Airtel combined those routines in ways that traditional telecoms had previously dismissed.

In Washington DC, two entrepreneurs chose to address one of the longest-running challenges in the American educational system: the failure of public high schools in the most impoverished and crime-ridden inner cities. They created an entirely new solution out of elements that had always been thought to be incompatible: high-cost boarding school practices, state-paid tuition, and locating in the target community. They created an entirely new organizational structure. That system linked The SEED Foundation, an overarching entity managed by business professionals, with boarding school entities, The SEED School of Washington DC and The SEED School of Maryland, managed by educational professionals. The SEED Foundation has been identified as one of the most innovative educational institutions in America, winning Harvard's Kennedy School Innovations in American Government Award.[14] Doing so put aside the routines associated with both educating urban youth and traditional boarding school education. SEED routinized administrative processes unknown to educational contexts as well as new educational activities, such as extensive postgraduation support services to the students going on to university. This routinization generated novel capabilities and systems that were fed back into the Foundation to support fundraising, community-level learning, and ultimately replication.

How much impact has SEED had? Beyond the students from poverty-stricken Anacostia who are graduating from college, beyond the economic impact in the local community, SEED's accomplishments have been recognized as changing how educators and politicians think about educating urban children. The landmark Edward M. Kennedy Serve America Act was signed on the SEED Campus in Washington DC. SEED co-founders Rajiv Vinnakota and Eric Adler were recognized by Oprah Winfrey's Angel Network as "Use Your Life" Award winners. But perhaps one of the most telling visits was that of Prince Charles and Lady Camilla in 2008 (see Figure 6.1), who found time between White House events to tour the DC School campus, sit in on classes, and talk to students:

> The 320-student school, the only public charter boarding school in the United States, was selected for a visit at the suggestion of Lady Catherine Manning, wife of the British ambassador [see Figure 6.1]. A representative of the British Embassy said the prince and the duchess were interested in taking ideas on expanding and improving such schools back to Britain, which already has a handful of them.[15]

4 *Develop new logic to widen horizons*

In extreme cases, entrepreneurs and innovative organizations employ seemingly new logic to widen the horizons of existing opportunities. At Return Path, Matt Blumberg and George Bilbrey turned the logic of spam filtering upside down. Rather than play to the innovative strengths of spammers, they saw a way to play to the strengths and budgets of legitimate email marketers and the ISP gateways. This has implications far beyond spam filtering. It has presented entirely new opportunities for improving email marketing. The company is in the early stages of implementing a much broader range of products, including Domain Assurance that protects against phishing

FIGURE 6.1 The SEED School of Washington, DC (top) being visited by President Obama (above)
Images courtesy The SEED Foundation

and spoofing attacks, which builds on the free authentication protocols like Domain Keys Identified Mail (DKIM) and Sender Policy Framework (SPF) that have remained ineffective at detecting and blocking on their own.[16]

Return Path's coherent story, which went against the dominant logic of the industry, was simply that fighting spam on its own terms was a losing battle. And so the company has shifted the ground of the battle to establish email sending standards. In the end, email marketers will benefit from more efficient and effective marketing campaigns, but everyday email users, for the most part unaware of Return Path's "center of the email universe" positioning, are likely to be the greatest beneficiaries.

Want to see what drives this sort of "un-logic"? Reading Matt Blumberg's blog on the Return Path website might reinforce the idea that innovative firms addressing novel opportunities have the option to apply truly original thinking that may not make sense to outsiders. Note, for example, Matt's "unpredictions" at the start of each year and his regular blogging about his interactions with his Board of Directors and even his performance reviews. How many CEOs operating in the spectacularly fast-moving software/email space would take a six-week sabbatical? You have to find a new logic if you want to expand your horizons.

5 *Deploy simple solutions for radical impact*

Perhaps the most powerful tool available to entrepreneurs is to deploy simple solutions for radical impact. At the time Google was spun out of Stanford University, the focus of the research effort was the deployment of a viable and plausible search heuristic for accessing information from internet content growing at hyperbolic rates. At the time, the primary goal was search effectiveness within a reasonable timeframe, given the exponentially increasing amount of data to be searched:

> Improving the performance of search was not the major focus of our research up to this point. The current version of Google answers most queries in between 1 and 10 seconds.[17]

But that would change. The challenge of relevance remains critical, especially in the context of SEO (search engine optimization) and unscrupulous content propagators who find ways to "game" Google's crawlers and cross-referencing heuristics. Some evolutionary modifications, such as Google Knol, seem plausible but fail to take hold. Others, such as the firm's recent efforts to ensure search outcome fidelity, are uncertain.[18]

But other, extremely simple-seeming mechanisms may have significant, even extraordinary impact. Here are two examples.

In the first, Google engineers poring over user data realized that users were slower than expected to click on the most likely link resulting from a search.

> Engineers noticed that while Google was returning useful pages when someone typed an acronym such as "CIA" – providing links to the government agency and to the Culinary Institute of America – people were taking a slightly longer time than expected to click on one of them. So on the results pages, Google began highlighting in bold the full names. Immediately, Google saw more clicks through to pages – and faster, too. How much faster? Perhaps 30 or 40 thousandths of a second, on average, [Google Research Fellow Amit] Singhal says. That's one tenth the speed of an eyeblink. "This was a small idea," concedes Singhal. "But we have a real responsibility as a company to respect people's time."[19]

Not fully convinced? In 2006, Marissa Mayer, Google's VP of consumer products, revealed the results of a so-called latency test. Google users have complained at times about the inherent limitations of only viewing ten results at a time. So Google ran experiments in which 30 results were displayed instead. The result? Users were a half-second slower to click on the preferred link, even when that link was in the top ten choices. Traffic, and revenue related to that traffic, dropped 20 percent. Imagine the effects of a 20 percent drop in revenue for a traditional manufacturing company because of a half-second delay in customer decision-making. In the interconnected, information-rich world, entrepreneurs and organizations have the potential to generate higher profits, but extremely small changes may have dramatic effects. Greg Linden, the developer of Amazon's recommendation engine, reinforces this effect:

> This conclusion may be surprising – people notice a half second delay? – but we had a similar experience at Amazon.com ... we tried delaying the page in increments of 100 milliseconds and found that even very small delays would result in substantial and costly drops in revenue.[20]

Craig Newmark also utilized a radical perspective to redraw the purpose of an online services organization. The extraordinary scaling power of the web meant that Craigslist customers, for extremely small incremental effort compared to a newspaper classified, could post classified advertisements for free. Yet the company doesn't charge either ad placements or ad viewers and doesn't monetize this resource with advertising. And Craig Newmark? According to the Craigslist.org factsheet, when he's not serving in his "iconic" role as founder, he does customer service. For Newmark, the simple solution was to break the rules of what a "for-profit" organization could define as its long-term goal. The impact on classified ad usage is impossible to ignore: members post more than 500 million ads each year on Craigslist, and the site exceeds 20 billion page views each year.

EMBRACING THE UNEXPECTED

These five actions: managing dissonance, unframing opportunities, routinizing to generate novelty, applying new logic to widen horizons, and developing simple solutions for radical impact, help entrepreneurs create previously unknown organizational forms.

These businesses are extremely hard to describe, as they often have no precedent. These are not cases of creative destruction, or discontinuous innovation – these are journeys into the unknown unknown. When entrepreneurs venture into these entirely unmapped spaces, the organizations they create are something new under the sun. They create hopeful monsters.

Hopeful monsters

Evolutionary biology once hypothesized the existence of organisms that deviated significantly from prior generations, presumably through randomly caused mutations. So-called *"hopeful monsters,"* coined by geneticist Richard Goldschmidt, were aberrations, unpredictable and spontaneous. Although genetic mutations might be relatively small, the effect on gene expression would be binary. To put it in familiar terms, a baby either has brown or blue eyes. A gene that suppresses certain cancer-related enzymes is expressed or not. Small gene expression changes would result in dramatic changes in physical characteristics, such as dinosaurs too heavy to support their own weight. These *monsters* were presumed to be unlikely to be well-adapted to the environmental niche of their parentage. But they were likewise *hopeful*, possibly well-adapted to a slightly different niche, perhaps a more resource-munificent niche, or a niche of increasing relevance due to broader environmental change. Evolutionary biology research shows that a vast majority of mutations are detrimental, especially in the case of organisms with highly complex genomes. But mutation always brings with it the potential to improve adaptation to the environment, especially in the context of environmental change or available resources.

Biology has generally rejected the "hopeful monster" hypothesis. But there is reason to believe that hopeful monsters exist in entrepreneurial contexts. First, the landscape of entrepreneurial firms is littered with unusual organizational forms and opportunity exploitation efforts that just didn't quite work: from ill-fated pursuits such as the DeLorean automobile, to illegal music file-sharing networks, and cross-industry ventures like Club Med Airlines. Some of these were doomed for structural reasons, others by poor implementation or mischance. And some were simply ill-timed: the Apple Newton wasn't much of a success, but its more sophisticated descendant, the iPhone, has revolutionized the mobile consumer communications industry.

Even more importantly, organizational forms that aren't well-adapted to the environment can do two things that individual organisms generally cannot. First, they can change. Second, they can change their local environment.

Entrepreneurs can grow and evolve organizations towards better coherence and environmental fitness. We have seen this at firms like CDI, Praj, Voxel, and Metalysis. The entrepreneurs who founded and run these companies designed their organizations to make opportunities accessible, turned imperfections to their advantage, remodeled the structures of their companies to reinforce coherence, and built bridges to new opportunities.

Even better, entrepreneurs can accomplish something more radical. They can change the nature of the opportunities they address, even change the environment in which their firms compete. The entrepreneurs at Return Path, Confederate, and Recurve have taken this more dramatic and innovative approach to opportunity exploitation.

These are the hopeful monsters, created in a context in which entrepreneurs refuse to be limited by resource constraints, goal constraints, process constraints, and sometimes even logic constraints. Entrepreneurial hopeful monsters simply are potentially adaptive agents of their own survival. Insight 4 provides the necessary tool for hopeful monsters to evolve: bridge-building. Insight 5 provides the tool to change the local environment: inspire the narrative. Entrepreneurs bringing radical innovations to market require something more than vision, implementation, resource acquisition skills, or passion. They use all five insights already discussed, *and* they embrace the unexpected.

Many entrepreneurs develop potentially brilliant innovations. But opportunities that deviate significantly from shared meanings in the status quo present greater challenges. In other words, opportunities that are high on the "monster" scale are presumed to be less well-adapted to the environment. Populations regress to the mean;

on the whole, we expect to see these effects play out in a business context. This is, after all, the basis for population ecology theories of strategy. Firms that balance exploration and exploitation have higher survival rates. Commercializing innovations requires "crossing the chasm" between early adopters and mainstream customers. Radical innovations require more convincing stories to acquire resources.

Hopeful monsters are radical, unexpected, and unpredicted – but they are often highly directed. While some innovations are spontaneous, many more are a response to information from the environment. Jerry Yang was cataloging the Internet for his own purposes and realized how valuable it was because everyone else needed it cataloged as well. Matt Chambers saw a coherent meaning that linked individual expression and industrial design. Matt Golden knew that many homeowners wanted to decrease their energy bills and improve their living environment. In other words, many hopeful monsters are designed, in part, to address an apparently available environmental niche.

Describing or explaining these entrepreneurs 20 years ago would have been restricted by the language that dominates corporate strategy in a competitive context. But entrepreneurs, by definition, find new ways to create and propagate meaning. The language of business models has provided that novel meaning-making framework. Using that new language, the new entrepreneurs design organizations to enact opportunities, rather than to achieve competitive advantage. The organization is the powerful tool that entrepreneurs leverage to bring opportunities into sharper focus, to propagate the vision of that opportunity's potential, and to structure organizational elements towards making the opportunity accessible.

During and after that structuring process, entrepreneurs perceive opportunities as shards rather than spheres. They acknowledge the inherent imperfections of opportunities, and appreciate that those very imperfections may be the key towards unlocking previously inaccessible value. Hopeful monsters are characterized by those imperfections. The imperfections of the opportunities, which may eventually be extraordinarily valuable, are described in traditional strategic language as uncertain or ill-fitting resources.

But entrepreneurs sometimes achieve the unexpected. They co-opt those imperfections, and remodel organizations for coherence. They link the imperfect elements of the organization and the opportunity into novel configurations that create entirely new meanings for the opportunity and the firm. In doing so, entrepreneurs converge towards coherent interpretations of the value-creating potential of a new opportunity. The coherent solution may not be optimized or even logical, but it makes sense within the organizational context, and provides meaning for the entrepreneur and the organization.

But these coherent solutions are only stable and valuable in the short term. Entrepreneurs that achieve the unexpected build bridges to span opportunities. At larger organizations, this requires simplifying structures to direct managerial attention outwards. Reconfiguring activities to improve internal efficiency actually makes the situation worse, because the organization effectively sacrifices opportunity searching for short-term gains. The challenge at larger organizations is even greater, because simplifying structures by ceding control of functions reduces the firm's ability to respond rapidly to change. Entrepreneurial managers implementing business model information must balance the need for managerial attention with the importance of retaining control.

Once bridges have been built, entrepreneurs must inspire the narrative that motivates crossing the bridge. In most cases we have observed, these narratives tend to derive from the original founding stories. But retaining the founding story too long can also be dangerous, if that narrative is made incoherent by exogenous change. We are narrative animals,[21] but the characteristics of storytelling familiar to us all are the same elements that enable managers to get locked into stories that have become dissonant. Entrepreneurial storytelling is an art enabled by extraordinary organizational talents and unique insight into human emotion and drive. We are only now beginning to appreciate what it may accomplish.

There are no long-term guarantees, as so many outlandish entrepreneurial ideas have demonstrated. But some of these hopeful monsters survive. Some of today's hopeful monsters will be tomorrow's Apple, Google, or Facebook.

Curiosity and wonder

> The truth is that there is no way to calculate whether money invested in business or money invested in helping to solve social problems will create more value ... To extend our love and care beyond our narrow self-interest is antithetical to neither our human nature nor our financial success. Rather, it leads to the further fulfillment of both. Why do we not encourage this in our theories of business and economics? Why do we restrict our theories to such a pessimistic and crabby view of human nature? What are we afraid of?
>
> John Mackey, founder of Whole Foods[22]

A theory of entrepreneurship that includes design, coherence, and embracing the unexpected complements much of what we know about entrepreneurs already, but helps explain much more.

One of the most important extensions addresses how entrepreneurs acquire valuable resources. Because entrepreneurship is highly idiosyncratic, clarifying the processes that entrepreneurs use to acquire strategically relevant resources has sometimes yielded conflicting results. Entrepreneurs create value from otherwise valueless resources in a process of bricolage. They narrate stories to external stakeholders to establish legitimacy via a process of cultural entrepreneurship. Some entrepreneurs run resource-lean entities and succeed, while others leverage slack resources to extend the organization.[23] Why do some entrepreneurs use one tool but not others? Are some of these tools more important than others, and if so, when, and why?

Our research points towards a more integrated framework for answering these questions. There is much work remaining, but we may take the first step. Innovative entrepreneurs apply an overarching heuristic to face an ill-formed opportunity or novel market space. They make sense of it.

This may seem obvious, but there are two less obvious outcomes of this theory. First, the sense-making process is not instantaneous. It is a convergence toward the best explanation in context. This is a significant departure from shared understanding of opportunity identification and even strategic decision-making more broadly. Entrepreneurs develop an understanding of what an opportunity means both through initial observation as well as the experiential outcomes of the exploitation process.[24] The convergence of the sense-making process may occur rapidly or slowly, and may or may not be confluent with the shared understanding at the organization. This brings us to the second important aspect of sense-making: it is specific to the individual. While some meanings are more universal and thus may facilitate broader shared interpretations, some may be highly idiosyncratic.

In the light of these outcomes, we suggest that the two dominant characteristics of entrepreneurs who embrace the unexpected are curiosity and wonder.

To be sure, there are likely minimum thresholds involved for a variety of other characteristics associated with more traditional entrepreneurship. We believe that curiosity and wonder come into play when entrepreneurs with a set of necessary skills and knowledge address the unknown unknowns. Many, perhaps most of them, choose to operate by reframing, focusing on what is familiar, and making the plausible possible. But some of the extraordinary entrepreneurs we studied were not content with this perspective. They brought curiosity, the desire to learn more for the sake of learning, as well as wonder, the childlike willingness to suspend disbelief in the near term for the benefit of longer-term, if uncertain, enlightenment.

In addition, they brought an inherent joy and optimism to the process, born of the belief that the journey was as valuable as the destination, and a willingness to be surprised, without anxiety regarding failure or being proven wrong.

The future of global opportunity

Like so many words overused in speech, media, education, and conversation, *opportunity* has become hackneyed and clichéd. It has become synonymous with potential, aspiration, and even democratic ideals. It has lost the mystique of uncertainty; it has become divorced from striving and challenge. In some contexts, opportunity has become *entitled*, something that should be parceled out fairly, in equal shares.

As fast as technologies change, social structures have evolved to adopt and propagate those technologies, sometimes for unexpected purposes. While generalizations must be made cautiously, it is undeniable that businesses such as Facebook or Twitter are playing unprecedented roles in international political theater. It is probably a testament to the perceived and real influence of these businesses that the Egyptian Government attempted to block access to the Facebook website during the January 2011 revolution that deposed the Government of Hosni Mubarak. The boundaries between technological, organizational, cultural, and political events have become fuzzier, overlapping, even intermingled.

The limitations of individual access, experience, and capability are being eroded.

Rough edges of opportunities are getting sharper. Even as more opportunities are available to more people, the nature of those opportunities presents more imperfections. Despite the concerns over Americanization of markets, and the rapid expansion of firms such as Wal-Mart and McDonalds, there are broader trends supporting specialization and differentiation. These may be superseded, for some time, in developing markets, where large, well-resourced firms leverage efficiencies in provisioning generic products and services. The leveling of access to opportunities hints at the long-term power of local entrepreneurship. It seems reasonable to expect that the multinationals that have developed in the past 20 years will find ways to localize products, services, technologies, and even organizational structures. But there are limits to growth, and ultimately, there are even limits to the growth of emerging economies. In those contexts, it falls to the entrepreneurs to identify where new opportunities may be created.

A variety of firms worldwide have begun to address markets emerging in the new context of ecological constraints and responsibility. This includes everything from sustainable office projects to carbon-neutral music festivals. Two of our study companies specifically focus on ecological outcomes: Metalysis and Recurve. Recurve, in particular, has developed a narrative of ecological sustainability that encompasses clients, partners, employees, and even its venture capital investors. These companies, and thousands of others, are addressing new and partially unproven opportunity landscape that has emerged from social and political trends associated with "green business."

It would be easy to disregard these efforts as faddish. Perhaps that is the case, but we think that such a perspective fails to take into account the changing nature of entrepreneurial activity. The advent of entrepreneurs into interconnected economies means that the impact of unsustainable investments and sustainable ventures will be known and incorporated into new opportunity recognition activities. The power of storytelling comes into play, because consumers are incredibly effective narrative interpreters.

Is this really happening, or is it just small-scale self-delusion?

If you live in the UK, you may be familiar with the Co-Operative, the fifth-largest food producing entity in the nation. The Co-Op is a mutual group, not a shareholder-based public company. This means that the customers are also the owners. And the customers at the Co-Operative have spoken. Executive management at the Co-Operative asked its members to help redefine the mission of the organization. Out of this process emerged visions for supporting renewable energy and social welfare. So the financial side of the business, Co-Operative Financial Services (CFS), put that in motion. In 1998 CFS established a renewable energies project fund. It began investing in the expertise to invest in relatively small-scale, embedded power generation facilities, especially in areas with high costs of energy transportation and infrastructure. It helped place a wind turbine on the island of Tyree in the Hebrides, which will save the island population of 700 residents at least £120,000 per year. And what will the island council do with that money? It's going to build a house for a secondary school science teacher.

But don't be fooled by this example, CFS is big business. The Co-Operative has a larger mortgage business than HSBC, and a larger business bank book than Royal Bank of Scotland. So when London's Natural History Museum, one of the world's largest and most prestigious public museums, needed an energy conservation retrofit, it turned to the Co-Operative. CFS has already invested £430 million in these types of projects, and is expanding its fund for these projects to £1 billion. And it is generating more profits than it ever has before.

THE BEGINNING

All great stories come to the end. But the story of new entrepreneurial activity around the world is just beginning. The economic crisis that began in 2008 slowed the growth of Indian technology and service companies, but didn't arrest it. The combination of the recession and the slow revaluation of the Chinese yuan have slowed economic growth in China, but not stopped it. The outcomes of political unrest in numerous countries at the time of this writing must have near-term effects on the globalized economy, and there are continuing indications that the full implications of the recession haven't been factored into the financial market's expectations.

There will be economic cycles, market corrections, and certainly unforeseen crises. One esteemed professor of finance and risk commented in his recent inaugural lecture that the only advice he could give about the next crisis was that it would most likely be in a sector we simply didn't expect.

It is for all these reasons, and all the reasons we have described throughout the book, that we believe that this is only the beginning of the impact of new entrepreneurial activity at all economic, industrial, and geographic levels. The population of the world will continue to increase, a few percent each year. Averaged out, global economic growth will tend to increase, a few percent a year.

But in the next 20 years, the number of entrepreneurs globally with access to resources, opportunities, and markets, will likely explode by a factor closer to ten than two. The most innovative of these individuals will have extraordinary impact on economic outcomes, social policy, and human welfare. We make no pretense to know exactly what they will accomplish nor suggest that there won't be failures along the way. But everything we have read, observed, and studied suggests that, on the whole, their work will be profitable for themselves, for society as we know it, and for the world. We expect nothing less from those who will achieve the unexpected.

NOTES

1 Drucker, P. 1985. *Innovation and Entrepreneurship*. New York: HarperCollins.

2 Baron, R. A. 1998. Cognitive mechanisms in entrepreneurship: why and when entrepreneurs think differently than other people. *Journal of Business Venturing*, 13: 275–294.

3 Palich, L. E. and Bagby, D. R. 1995. Using cognitive theory to explain entrepreneurial risk-taking: Challenging conventional wisdom. *Journal of Business Venturing*, 10(6): 425–438.

4 Hmieleski, K. M. and Baron, R. A. 2008. Regulatory focus and new venture performance: a study of entrepreneurial opportunity exploitation under conditions of risk versus uncertainty. *Strategic Entrepreneurship Journal*, 2(4): 285–299.

5 Sarasvathy, S., Simon, H., and Lave, L. 1998. Perceiving and managing business risks: differences between entrepreneurs and bankers. *Journal of Economic Behavior and Organization*, 33: 207–225.

6 McGrath, R. G. and MacMillan, I. C. 2000. *The Entrepreneurial Mindset: Strategies for Continuously Creating Opportunity in an Age of Uncertainty.* Boston, MA: Harvard Business School Press.

7 Heavey, C., Simsek, Z., Roche, F., and Kelly, A. 2009. Decision comprehensiveness and corporate entrepreneurship: the moderating role of managerial uncertainty preferences and environmental dynamism. *Journal of Management Studies*, 46:1289–1314.

8 Argyris, C. 1977. Organizational learning and management information systems. *Accounting, Organizations and Society*, 2(2): 113–123.

9 Cohen, W. M. and Levinthal, D. A. 1990. Absorptive capacity: a new perspective on learning and innovation. *Administrative Science Quarterly*, 35(1): 128–152.

10 Zahra, S., Ireland, D., and Hitt, M. 2000. International expansion by new ventures: international diversity, mode of entry, technological learning and performance. *Academy of Management Journal*, 43: 925–950.

11 Mitchell, R. K., Busenitz, L., Lant, T. *et al.* 2004. The distinctive and inclusive domain of entrepreneurial cognition research. *Entrepreneurship Theory & Practice*, 29: 505–518; Weick, K. E, Sutcliffe, K. M., and Obstfeld D. 2005. Organizing and the process of sensemaking. *Organization Science*, 16(4): 409–421.

12 George, G. and Bock, A. J. 2010. The role of structured intuition and entrepreneurial opportunities. In: Phillips N., Griffiths, D., and Sewell, G. (eds), *Technology and Organization: Essays in Honour of Joan Woodward (Research in the Sociology of Organizations, Volume 29).* Oxford: Emerald, pp. 277–285.

13 Haynie, J. M., Shepherd, D. A., Mosakowski, E., and Earley, C. 2011. Cognitive adaptability: metacognition and the "entrepreneurial mindset." *Journal of Business Venturing*, 25(2): 217–229.

14 Source: www.innovations.harvard.edu/awards.html?id=7498 (accessed March 9, 2011).

15 "Royal couple find friends at White House and a school" by Marian Burros, *New York Times*, November 3, 2005, www.nytimes.com/2005/11/03/politics/03royals.html (accessed July 1, 2011).

16 Source: www.Return Path.net/blog/intheknow/2011/02/return-path-research-shows-trusted-brands-are-wide-open-to-phishing-attacks-domain-assurance-can-change-that/ (accessed February 28, 2011).

17 Brin, S. and Page, L. 1998. The anatomy of a large-scale hypertextual Web search engine. *Computer Networks and ISDN Systems*, 30(1–7): 107–117.

18 "Google's search cleanup has big effect" by Amir Efratie, *Wall Street Journal*, February 28, 2011. www.online.wsj.com/article/SB10001424052748704288304576170851535102540.html (accessed March 2, 2011).

19 "Can Google stay on top of the web?" by Robert D. Hof, *Bloomberg Businessweek*, October 1, 2009. www.businessweek.com/magazine/content/09_41/b4150044749206.htm (accessed March 2, 2011).

20 Source: www.glinden.blogspot.com/2006/11/marissa-mayer-at-web-20.html (accessed February 27, 2011).

21 Fisher, W. R. 1987. *Human Communication as Narration: Toward a Philosophy of Reason, Value, and Action.* Columbia: University of South Carolina Press.

22 "Rethinking the social responsibility of business," *Reason* Magazine, October 2005.

23 Mishina, Y., Pollock, T. G., and Porac, J. F. 2004. Are more resources always better for growth? Resource stickiness in market and product expansion. *Strategic Management Journal*, 25: 1179–1197; George, G. 2005. Slack resources and the performance of privately held firms. *Academy of Management Journal*, 48(4): 661–676.

24 Krueger, N. F. 2007. What lies beneath? The experiential essence of entrepreneurial thinking. *Entrepreneurship Theory & Practice*, 31(1): 123–138.

Hindsight

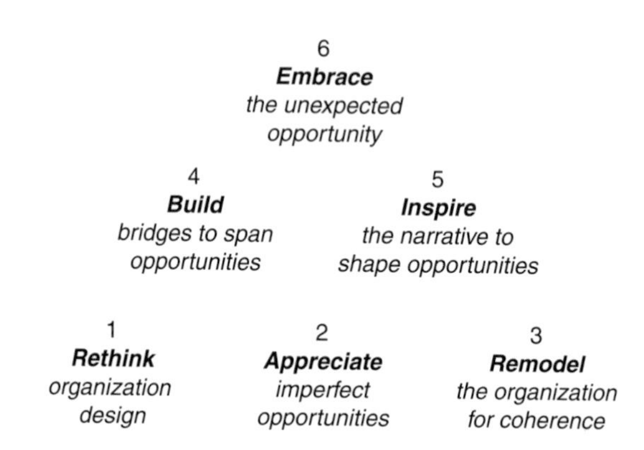

Achieving the unexpected is not the norm for entrepreneurial ventures.

In a senior leadership program held in India, a group of European CEOs were keen to understand why business models work differently in India. One participant noted wryly, "Everything we try [on an Excel sheet] tells us this is not an opportunity. And yet an Indian entrepreneur takes it and makes it work brilliantly. I am just not sure how to do business in this context. What is the trick that I am missing?"

Indian or Chinese entrepreneurs do not have better business model recipes. But because many of them don't bring Western "common sense" to their ventures, they don't accept the boundary conditions of plausibility self-imposed by many Western managers based on traditional strategic frameworks. But this book is not intended to convince CEOs most comfortable with organizational precision to remodel their organizations for coherence and embrace the unexpected. For now, translating our insights into a spreadsheet to make unusual opportunities seem credible remains difficult or impossible. It is possible that, in the future, further research will help reveal the underlying mechanisms and clarify the predictive heuristics for when organizational coherence makes the difference between a failed European start-up and a rapidly growing Chinese business.

We are continuing our research in this direction. We are studying how entrepreneurs first perceive and make sense of opportunities long before forming a legal business, and we are initiating research on the contextual factors that drive narratives of opportunity assessment. We are hopeful that psychological constructs such as fluency and sociological frameworks such as cultural narratives will help inform our understanding of entrepreneurial activity. A science of organizational coherence is still in the future; predicting the apparently unpredictable is beyond our current capabilities.

Although our analysis incorporates a relatively global perspective, much of the truly revealing information came from decidedly Western firms. It would be naïve to anticipate that the lessons of what are predominantly *Western* stories are immediately and substantively applicable to organizational narratives of decidedly *non-Western* entrepreneurs and businesses in the largest emerging economies.

Perhaps the formalization of our insights will require a more Eastern-oriented philosophy, in which knowledge and understanding are not limited to rational, sequential analysis, but are sometimes arrived at via moments of unexpected clarity. It may seem unlikely that business schools would teach corporate strategy with *koans*, Zen stories of paradox that have no rational meaning or solution. On the other hand, 30 years ago it seemed unlikely that industry attractiveness could be assessed quickly with a simple diagram, yet now we often teach Porter's Five Forces model to MBAs in only a few minutes. Perhaps someone will deconstruct a strictly analytical, empirical process for our six insights.

Today's entrepreneurs and executives face unprecedented challenges. Although endearing flaws, design for coherence, and bridging narratives are powerful tools, the six insights won't work for all of us all of the time. These insights do not provide a recipe for success, nor do they represent an action plan for most entrepreneurs. Coherence is not a panacea for survival, nor does it convert temporary competitive advantage into sustainable competitive advantage. While Confederate Motorcycles has defied the apparent limitation of operational management and evolved through local economic sinkholes and global economic earthquakes, Savage Entertainment functioned with a coherent if dissonant story for a decade but couldn't bridge the opportunity gap when industry dynamics changed. For now, at least, organizational coherence is, at best, a trigger – a trigger to consciously address plausibility and to push the boundaries of what makes an opportunity work. The extraordinary entrepreneurs among us will take an opportunity that looks implausible and make it real.

Looking back, we want to emphasize why the six insights are not relevant for all entrepreneurs, and to suggest some indications for when they present the most value.

The vast majority of entrepreneurial businesses are very small, non-technology businesses. In the United States there are more than 10 million businesses with less than 50 employees, mostly sole proprietorships. Few, if any, of these organizations will achieve dramatic, unexpected results. They are companies that generate a sustainable living for founders and a few employees, often family members. Roughly half the workforce in the United States is employed by large corporations. Some of these organizations have achieved unexpected results in an early growth phase, and many could use the six insights to identify and effectively target novel opportunities. Few, however, will achieve such unexpected outcomes repeatedly.

It is an indicator of the unique time in which we write this book that some of the familiar large companies do continuously achieve unexpected outcomes. One is Apple's successive introductions of the iPod, iTunes, iPhone, and iPad, transforming the firm from a computer company to a provider of information-based experiences.

Another appears to be Google, leveraging its enormous data infrastructure into a mobile operating system and pushing boundaries in other directions ranging from sustainable venture capital to automobiles that drive themselves. To be sure, neither of these companies has, nor will, achieve extraordinary results all the time: witness Google Wave and Apple's current iPhone 4.0 antenna problems. Continuous, repeated success in achieving the unexpected is, by far, the radical exception in the world of large corporations. Research suggests that business model innovation has become essential to continuous outperformance, especially beyond the short term.

This should be worrying to senior managers because it appears that successful business model innovation is not linked to prior learning or success. In other words, we don't know if business model innovation can be taught, learned, or routinized. In contrast, it appears that some firms are finding ways to make business model innovation work repeatedly: perhaps Apple via extraordinary insight, Google via broad and distant opportunity search. It is interesting to note that these very exceptions have some of the most compelling organizational stories, narratives that capture our imagination.

For all these reasons, and many more, the lessons of highly innovative, exceptional organizations should be applied with caution in a more general context.

WHEN TO USE THE SIX INSIGHTS

The six insights have greatest relevance in a highly specific context. Putting aside geographical and cultural issues, we can suggest the type of organizations and circumstances that have the greatest potential to leverage the six insights.

First, firms that address entirely new opportunity sets should think about Insights 1, 2, and 3 to approach opportunities with a focus on organizational design. Early stage companies addressing novel or speculative opportunities may suffer from imprinting effects and biases that place non-obvious limitations on capabilities. This may be especially true when businesses exploit opportunities that are ill-formed or uncertain. Cognitive psychology research shows that we tend to subconsciously apply reference frames to unsolved problems. The more unfamiliar the problem, the more we search for familiarity. It is precisely when CEOs have the most difficulty understanding an opportunity that they perceive those opportunities to be implausible and inaccessible. The very process of contextualizing the unknown becomes counterproductive, because it restricts the entrepreneur's ability to generate a new and plausible story. Mapping unknown variables into a known story may constrain entrepreneurial behavior. The most powerful stories, and the mechanisms for shaping organizations with those stories, benefit from coherence, not familiarity.

Second, many established firms seek to broaden opportunity sets and transition from one opportunity to another. Entrepreneurial managers at these firms may find Insights 3, 4, and 5 helpful to encourage new thinking and explore previously inaccessible opportunities. These are especially relevant when firms, large or small, must prioritize and enact tradeoffs between opportunities. This process is difficult, as most managers have learned, because prioritizations are tightly bound to organizational narratives. Moving the organization to new opportunities, and ultimately to new competitively advantageous positions, may require changing the narrative in concert with adjusting priorities and activities. Maintaining coherence is essential, precisely because we are all experts at interpreting stories. We are, as Fisher has noted, "narrative animals," from the youngest child to the most experienced CEO. We know good stories when we hear them, and we spot flaws in bad stories with equal facility. Insights 4 and 5 build on this to show how firms rewrite stories to inspire and motivate, while Insight 3 provides the touchstone for solidifying the organization around coherent narratives.

Large and small firms achieve coherence via different means. Large firms rely heavily on structural artifacts and process systems. Leadership and vision are powerful tools, but any organization of reasonable size must utilize structuring mechanisms in change processes and propagating meaning across the organization. The IBM data shows the importance of structural simplification to focus managerial attention. Business model innovation requires balancing simplification with control while narrating change that makes sense to stakeholders. Larger organizations are poly-vocal; the variety of stakeholder stories must be channeled into the dominant organizational narrative for change and opportunity search.

Examples of dramatic narrative change at large organizations have become more evident. When Ratan Tata galvanized support for the Nano car, he invoked narrative to establish a beachhead and guided Tata Motors to completely redesign the organization for the opportunity, right from supply chain to shop floor and distribution. At that time, Tata Motors did not have the capability nor the cost structure to make the opportunity seem sensible. Yet, Tata achieved the unexpected through narrative coherence, organizational redesign, and a dollop of inspiration.

Coherent narratives at large organizations may carry extraordinary inertia. It would be difficult to overstate Steve Jobs' role in Apple's rise to dominance in the consumer technology space. The media attributes the successes of the organization to his inspiration, the failures to poor implementation. One of the greatest challenges, then, for the multinational CEO, is to understand when new narrative coherence requires drastic action. Lou Gerstner taught the IBM elephant to dance, but in reality IBM's image as white-shirted conservatives was outdated by that time. The technological knowledge and market savvy at the world's dominant information technology firm were dormant, waiting for the narrative authority to emerge. Not entirely different, Nokia CEO Stephen Elop's brutally honest "burning platform" memo may or may not work to the company's long-term survival. Building coherent organizational narrative at the multinational organization remains complex and poorly understood. We don't know whether firms can be shocked through narrative change to a new coherent state. But we do know that even the largest organizations can build bridges to new opportunities.

Small firms have much more flexibility – founders and key managers exert narrative influence rapidly and affirmatively across the entire organization. Small, innovative firms tend to be univocal; some sound like a monologue spoken by one inspiring entrepreneur. These firms, however, often bring fewer capabilities and significantly less experiential knowledge to enact a new opportunity. For these entrepreneurs, the challenge lies less in coordinating the interpretation of the narrative, and more in justifying and empowering the narrative of change with a limited base of resources.

The six insights do not comprise a recipe for commercial success. They provide guidance and reflection on the unique capabilities of entrepreneurs who leverage the most fundamental drivers of human behavior – belief and aspiration. And just as this capacity may support extraordinary accomplishment, it can also lead to exemplary error. There is little doubt that a narrative of land-grabbing speculation and optimism influenced the events and activities leading to the dot-com bubble, where the unexpected led to significant financial losses for many firms and investors.

Silicon Valley and places like it embody powerful narratives that capture our imagination and harness entrepreneurial potential. Managers, employees, entrepreneurs and financiers believe that incredible and unexpected things can happen. This is the narrative that funded a search engine based on an algorithm for academic citation analysis, and a start-up social network with no monetization plan. Google and Facebook did not succeed *because* of the Silicon Valley narrative. But their success was *empowered* by that narrative, which places a special value on how entrepreneurs *make sense* of opportunities, rather than just whether the opportunities make sense.

A FEW UPDATES

Books on business management, especially books on entrepreneurship, are generally out-of-date even before they publish. Although we believe much of the theory and framework in this book will prove useful for years, our printed storytelling can't match pace with the entrepreneurial storytelling in the real world.

Just as we finished the first full manuscript of the book in Spring 2011, the final chapter in the Savage Entertainment story was told. You may recall the landscape shift in the gaming industry that resulted in Savage dramatically scaling back its operations. Founder and CFO Chacko Sonny returned to Activision soon after. Ultimately, what remained of Savage was sold to Loyalize, a company focused on social television applications. Founder and CEO Tim Morten oversaw the acquisition and then moved on to Victory Games, an Electronic Arts studio in Los Angeles, where he heads up development of the next *Command & Conquer.*

Broadjam continues steady expansion. The company serves more than 100,000 website users covering 190 countries. It continues to provide music catalog consulting work for one of the largest music publishers in the world. And each month it helps its customers license almost a hundred new songs into movies, TV, advertisements, and even video games like *DanceDanceRevolution.*

Cellular Dynamics (CDI), the stem cell company spun out of the University of Wisconsin-Madison, raised a further $30 million in venture funding, bringing its total venture funding to roughly $100 million. In 2011 it was named one of the 50 most innovative companies in the world by MIT's Technology Review.

Return Path continues its epic mission to make email safer, more reliable, and more secure. In July 2011, the company announced that its Feedback Loop Service, which helps ISPs manage email complaints, manages and hosts ten of the world's 13 public complaint feedback loops. The company has continued its rapid expansion. While the company just hired its 300th employee in Fall 2011, it plans to hire 100 more in 2012. And perhaps it comes as no surprise that *Inc.* Magazine named Return Path one of the top 20 workplaces in the United States.

But what about our cliffhanger story? Were we right about Voxel.net?

Yes and no. We were more right than we even realized when we drafted the first version of this manuscript. The team at Voxel.net had been struggling with the question of scale-up and the opportunity chasm before them. After we completed our interviews at the company, the founders, Raj Dutt and Zachary Smith, had a series of heart-to-heart meetings to discuss their options. They had arrived, completely independently, at very similar conclusions: they had to grow to compete, or they risked the firm being relegated to second-tier status. A limited number of options was on the table: selling or even shuttering the business, having one of the founders take over while the other carefully ramped down involvement, or finding an unexpected solution. That solution emerged in late 2010, when the company began meeting with Raul Martynek, a successful telecom entrepreneur and executive who had targeted the cloud space for his next venture.

Voxel's challenge was that it couldn't raise traditional venture capital on a relatively low EBITDA business model because it couldn't demonstrate an executive track record of scaling, but it couldn't scale without venture capital. The company had been in tentative and slightly uncomfortable discussions with a few venture financing groups. Instead, the firm crafted a preliminary arrangement with Martynek that led to more targeted efforts to identify active, industry savvy investors. Martynek officially joined the company as CEO in February 2011, and a $5.5 million investment from Seaport Capital followed in late March. Seaport is an early-mid stage ITC investment specialist with an active portfolio and an extensive history of successful scale-ups and exits.

Was this the skyhook Voxel.net needed? We're not sure. The challenges we identified still exist. But it was certainly a significant step forward. CEO Martynek and founder Zachary Smith, reflecting on this question, commented that the skyhook might have been the combination of the founders' willingness to take on "adult supervision" at precisely the one time that the market was willing to fund a complete rethink on organizational goals, structures, and resources.

Perhaps it was just that the potential bridge couldn't be seen from where the founders were standing. The picture of the smiling team is still up on the website, and perhaps that's enough for now.

CAUTIOUS OPTIMISM

As authors, we are, perhaps, unusual communicators for this set of insights. First, we bring a sense of caution inherent to the academic research process. We are keenly aware of both the strengths and weaknesses of our research. We believe there is much our research has to offer to both the science of entrepreneurship research and the art of entrepreneurship practice. At the same time, there is far more that we don't know, and we must be cautious in presenting observations as anything more definitive than insight.

On the other hand, we also speak to the power of entrepreneurial action from personal experience. Although we have both chosen to focus our careers primarily in academic settings, we are no strangers to the venturing world. Even as this manuscript was drafted, one of our firms executed a Series A round of venture capital. Another tripled its revenues, and a third was acquired by a publicly traded firm. These three businesses, in robotics, renewable energy, and biomedical devices, span dramatically different business models, corporate cultures, and 13,000 kilometers of geography. While it is unlikely any will become household names, or even industry leaders, each has brought an unlikely opportunity closer to reality. We ourselves have participated, in some cases unknowingly, in the creation of coherent narratives, some more successful than others.

We are optimistic that entrepreneurship presents a potent and efficient instrument for social change, economic development, and improvement of the human condition. The efforts of large organizations that contribute positively to society in addition to returning profits to shareholders should be celebrated. There is little doubt that firms like Wal-Mart, BP, SinoPec, Infosys, and Tata have the power to influence social and government policy for global good. At one time, it was accepted wisdom that corporations were blind to or disinterested in the broader implications of their business practices. The interconnectedness of social and organizational spheres is no longer in doubt.

We also believe that extraordinary and discontinuous change will primarily emerge at the local level. These changes arise from the cumulative efforts of new cohorts of entrepreneurs with decidedly different priorities and the unique effect of rare, explosive growth businesses that change the very fabric of the business environment. Companies like Google and eBay have changed fundamental characteristics of doing business on a global scale; Facebook has become a critical communication mode for organizations, individuals, and political movements. These enabling systems have empowered local entrepreneurs to extend impact beyond the local scale.

The drivers of change, whether in mature economic clusters like Silicon Valley or in rapidly evolving geographies like Mumbai, Shenzhen, and Tel Aviv have begun to incorporate more globally relevant factors. With a few notable exceptions, the communities of the world are interconnected, with information available on a moment's notice, 24/7. It is interesting, in this context, to ponder the observation effect, when it is not just a line manager or a researcher watching, but the entire world, all the time. One conclusion would be paralysis, fear, and paranoia. But we believe that what is actually emerging, in a testament to human resilience and the narrative of human learning and growth, is an emerging understanding and interpretation of that interconnectedness. The world has changed, and, optimistically, we choose to see these changes for the better.

Not every community can be Silicon Valley. We think, however, that its uniqueness will ultimately fade somewhat as other cities and regions flourish. The next generation of Silicon Valleys will perhaps be equally brave in funding coherent but outrageous organizations. And they hold great promise for entrepreneurs achieving the unexpected, and making the world a better place at the same time. Despite its lure, we wouldn't all choose to live in Silicon Valley anyway.

Finally, an important trend is the subtle shift of entrepreneurial vision from profit maximization or financial independence to survival and social purpose. More and more of the innovative entrepreneurs that we spoke to focus on what the business can accomplish rather than how much personal wealth they will generate. Very often those accomplishments incorporate an appreciation for obligations and responsibilities at the community level. This type of venturing has recently been labeled "social entrepreneurship" – a categorization that is arguably counterproductive because entrepreneurship is inherently a social process. Though entrepreneurship has at times generated spectacular monetary gains, very often even these examples have also created phenomenal communal goods. In just one example, we benefit significantly as academics from the Google Scholar tool, which provides one of the most comprehensive systems cataloging much of the world's scholarly knowledge. There may be no such thing as a free lunch, but there are externalities and spillover effects from entrepreneurial activities that present previously unavailable benefits on a grand scale.

The last 25 years of entrepreneurship scholarship have yielded a wealth of knowledge about entrepreneurship. We know some of the broader indicators of entrepreneurial activity, and some of the most common drivers of success. But much of entrepreneurial action remains imperfectly characterized. For all our efforts, many key questions are unanswered. Why do some people become entrepreneurs but not others? Why do some entrepreneurs see certain opportunities? Which skills are the most important for success in entrepreneurial venturing? Why do some ventures succeed when others fail? We have partial answers to these questions, but for every macro-level trend supported by statistical research, we find outliers and counterexamples.

Perhaps we shouldn't be surprised or even disappointed. After all, many of the greatest entrepreneurial successes, including Baidu, Facebook, Google, Genentech, Reliance, Tata, Apple, and Virgin, represent fundamentally unexpected achievements. The fastest growth businesses present stories of firms that succeeded where most others did not. In other words, the companies that we attempt to model are outliers. It is always difficult to develop theories for such outliers, which is why we tender our observations with due caution.

For the moment, we cannot provide a recipe of actions that guarantee extraordinary results. After all, such a prescription would seem to necessitate recalibrating what extraordinary would mean. Large cohorts of highly competitive and innovative companies would change the dynamics of industries and markets. Systematically raising the bar is inevitable, but in the end not all firms can be exceptional. Business is rarely a zero-sum game, but macro-level economies have never grown as rapidly as outlier firms within those economies.

This may be part of the driving force behind propagation of other organizational forms, such as highly innovative entrepreneurship in private, public, and quasi-public sectors. These organizations do not define long-term success primarily as extraordinary profits. Their narratives thus incorporate growth horizons with more space to expand without encountering traditional forms of competition. As the boundaries between for-profit firms and non-traditional forms become more blurred, even this may change. In the near term however, these ventures foster some of the most innovative entrepreneurs, including the founders of EMRI, JaipurKnee, The SEED Foundation, and many others.

We have come to the end of this story. We depart with a shared sense of wonder for the organizations and entrepreneurs that will lead new economies, invent new technologies, provide new services, and create new markets. Near-term economic challenges and emerging uncertainty in certain political theatres appear to present boundaries and limitations. But we are optimistic that the coming decades will bring entrepreneurial activity on a scale previously unimagined. It is a time of unparalleled opportunity for entrepreneurs.

Index